BLACK HOLE WINDS AT ALL SCALES

IAU SYMPOSIUM 378

COVER ILLUSTRATION:

Artistic representation of an active galactic nucleus winds
Credit: NASA/JPL-Caltech https://www.jpl.nasa.gov/images/pia18919-how-black-hole-winds-blow-artists-concept

IAU SYMPOSIUM PROCEEDINGS SERIES

INTERNATIONAL ASTRONOMICAL UNION
UNION ASTRONOMIQUE INTERNATIONALE

International Astronomical Union

IAU

BLACK HOLE WINDS AT ALL SCALES

PROCEEDINGS OF THE 378th SYMPOSIUM OF THE INTERNATIONAL ASTRONOMICAL UNION HAIFA, ISRAEL 12–16 MARCH 2023

Edited by

GABRIELE BRUNI
INAF, Italy

MARIA DIAZ TRIGO
ESO, Germany

SIBASISH LAHA
NASA – Goddard Space Flight Center, United States

and

KEIGO FUKUMURA
James Madison University, United States

CAMBRIDGE
UNIVERSITY PRESS

CAMBRIDGE UNIVERSITY PRESS
University Printing House, Cambridge CB2 8BS, United Kingdom
1 Liberty Plaza, Floor 20, New York, NY 10006, USA
10 Stamford Road, Oakleigh, Melbourne 3166, Australia

First published 2024

Printed in Great Britain by Henry Ling Limited, The Dorset Press, Dorchester, DT1 1HD

Typeset in System LaTeX 2ε

A catalogue record for this book is available from the British Library of Congress Cataloguing in Publication data

ISBN 9781009398817 hardback
ISSN 1743-9213

Table of Contents

Chapter 3: Feedback on the host galaxy

Chapter 4: Outflows from stellar black holes

Chapter 5: Theory & Modelling of outflows

Preface

Black holes are observed to drive powerful outflows on both stellar and galactic scales. The connection of the outflow to the accretion flow that feeds the black hole remains contentious. The physical launching mechanisms and astrophysical sites from which these outflows are launched also remain in debate; radiation, magneto-hydrodynamics, and thermal driving have all been proposed and modeled, with no consensual conclusion. Beyond their physics, the study of outflows is important for understanding how they affect their environment and host galaxy. The means by which outflows interact with their surroundings manifests itself in all wavelengths. Some examples of which are X-ray and UV ionized absorbers and obscurers, ultra-fast outflows at a fraction of the speed of light, massive molecular outflows, and extended radio emission. Despite significant progress in multi-wavelength observations, a coherent picture of black hole outflows has yet to emerge.

The goal of this IAU symposium was to better understand black hole winds by seeking common physical grounds for different outflows, in both stellar (X-ray binaries) and AGN (active galactic) black holes, in all wavelengths, and in the context of competing theoretical models. The connection of the winds to star formation, galactic feedback, and relativistic jets were discussed among observers, modelers, and theoreticians.

Seventy astrophysicists, both experts and students, convened for five full days at the Technion campus in Haifa. The exciting and popular topic brought together observers of many different wavebands, which made for a multitude of new data being presented and discussed. There was a good mix of low-energy and high-energy observations, including ground facilities and space observatories. Theoretical aspects of black hole outflows were also a big part of the discussions, and perhaps as expected, many puzzles about these winds remain.

In terms of new observations, IR astronomers presented JWST data indicating significant galactic feedback at redshifts above $z = 6$. New insights into the central sub-pc regions of active galaxies from GRAVITY were presented. In the mm-band, rich ALMA spectra covering a broad band provided new evidence that even the most obscured sources may host an accretion disk and produce outflows. Several new VLBA observations showed that even the radio quiet quasars host a compact, optically thick radio source at their core, while some of them also feature extended optically thin emission that can be associated with an outflow. In the UV with Hubble, broad absorption line systems allowed for quantitative estimates of mass outflow rates indicating that massive outflows in the most luminous quasars carry appreciable mass and energy out to kpc scales in the galaxy.

On the theoretical ground, a variety of simulations for magnetic and thermally driven winds were presented, some of which attempted to put galactic and stellar black hole winds on an equal footing. Cases from extreme mass accretion rates to advection dominated flows were covered. Topics of radiative driving and radiative transfer in outflows were discussed, a few of which even proposed comparisons with observations.

This volume collects contributions on the different themes discussed during the symposium, and in particular: the multi-wavelength picture of AGN outflows, spectroscopic methods, feedback on the host galaxy, outflows from stellar mass black holes, and theory & modelling of outflows.

Ehud Behar
Gabriele Bruni
(SOC co-chairs)

Editors

SOC information

- Ehud Behar, Technion – Israel Institute of Technology, Israel (co-chair)
- Gabriele Bruni, INAF, Italy (co-chair)
- Yair Krongold, Universidad Nacional Autonoma de Mexico, Mexico (co-chair)
- Anna Lia Longinotti, Universidad Nacional Autonoma de Mexico, Mexico (co-chair)
- Susanne Aalto, Chalmers University, Sweden
- Maria Diaz Trigo, ESO, Germany
- Christine Done, University of Durham, United Kingdom
- Keigo Fukumura, James Madison University, United States
- Sibasish Laha, NASA – Goddard Space Flight Center, United States
- Jon Miller, University of Michigan, United States
- Raffaella Morganti, ASTRON, Netherlands
- Gabriele Ponti, INAF, Italy
- Daniel Proga, University of Nevada, United States
- Francesco Tombesi, Università di Roma Tor Vergata, Italy
- Alexander Wagner, University of Tsukuba, Japan

LOC information

- Keren Tshilibi, Technion – Israel Institute of Technology, Israel
- Ayala Hodara, Technion – Israel Institute of Technology, Israel
- Adi Levy, Technion – Israel Institute of Technology, Israel

List of Participants

Name	Surname	Affiliation
Ajay	Ratheesh	INAF-IAPS
Aldana	Grichener	Technion
Anna	Ogorzalek	Stanford University
Anna Lia	Longinotti	UNAM
Ari	Laor	Technion
Asaf	Pe'er	Bar Ilan Univeristy
Assaf	Horesh	Hebrew University of Jerusalem
Aylecia	Lattimer	University of Colorado Boulder
Bartolomeo	Trefoloni	Arcetri Astrophysical Observatory
Chentao	Yang	Chalmers
Chiara	Feruglio	INAF
Christian	Partmann	Max Planck Institute
Christopher	Harrison	Newcastle University
Clare	Wethers	University of Turku
Claude-Andre	Faucher-Giguere	Northwestern University
Cosimo	Marconcini	Arcetri Astrophysical Observatory
Daisuke	Toyouchi	The University of Tokyo
Deborah Kala	Perkins	InfinitiEd
Dieter	Lutz	Max Planck Institute
Doron	Chelouche	University of Haifa
Ehud	Behar	Technion
Fabrizio	Fiore	INAF-OATrieste
Fangzheng	Shi	Shanghai Astronomical Observatory
Feng	Yuan	Shanghai Astronomical Observatory
Francesca	Panessa	INAF-IT
Francesco	Massaro	Turin University
Francesco	Bollati	University of Milan
Gabriela	Calistro Rivera	European Southern Observatory
Gabriele	Bruni	INAF - Institute for Space Astrophysics and Planetology
Giovanni	Cresci	Arcetri Astrophysical Observatory
Hagai	Netzer	Tel Aviv University
Hyunseop	Choi	University of Montreal
Igor	Gitelman	Technion
Ivan	Katkov	NYU Abu Dhabi
Jonathan	Stern	Tel Aviv University
Joseph	Gelfand	NYU Abu Dhabi
Kirsty May	Butler	IRAM
Labani	Mallick	Caltech
Lorenzo	Ulivi	Arcetri Astrophysical Observatory
Marcella	Brusa	Universiy of Bologna
Marco	Alban Morales	Heidelberg University
Marina	Bianchin	UC, Irvine
Marzena	Sniegowska	Tel Aviv University
Matthew	Temple	Universidad Diego Portales, Chile

Matthew	Stepney	University of Southampton
Maxim	Lyutikov	Purdue University
Mukesh Kumar	Vyas	Bar Ilan Univeristy
Nahum	Arav	Virginia Tech
Navin	Sridhar	Colombia University
Neeraj	Kumari	Physical Research Laboratory, India
Nick	Kylafis	Institute of Astrophysics, FORTH
Nina	Akerman	UNIPD
Noa	Keshet	Technion
Paola	Rodriguez Hidalgo	University of Washington
Pierre Olivier	Petrucci	IPAG/CNRS
Sarath	Satheesh Sheeba	Universidad Andres Bello, Chile
Sharon	Miterani	Technion
Sina	Chen	Technion
Steve	Kraemer	Catholic University of America
Sudip	Chakraborty	CEA Paris-Saclay
Susanne	Aalto	Chalmers University
Thong	Le	Ton Duc Thankg University, Vietnam
Victor	Oknyansky	University of Haifa
Vincenzo	Mainieri	European Southern Observatory
Vittoria Elvezia	Gianolli	Université Grenoble Alpes
VIVEK	M	Indian Institute of Astrophysics
Wang	Sili	University of Georgia
Yair	Krongold	UNAM
Yuanze	Luo	Johns Hopkins
Maria	Díaz Trigo	ESO

Chapter 1:
The multi-wavelength picture of AGN outflows

Black Hole Winds at All Scales
Proceedings IAU Symposium No. 378, 2024
G. Bruni, M. Diaz Trigo, S. Laha & K. Fukumura, eds.
doi:10.1017/S1743921323003514

Interpretation of IR variability of AGNs in the hollow bi-conical dust outflow model

Victor L. Oknyansky[1,2] and **C. Martin Gaskell**[3]

[1]Department of Physics, Faculty of Natural Sciences, University of Haifa, Haifa 3498838, Israel
email: victoroknyansky@gmail.com

[2]Sternberg Astronomical Institute, M.V. Lomonosov Moscow State University, 119234, Moscow, Universitetsky pr-t, 13, Russia

[3]Department of Astronomy and Astrophysics, University of California, Santa Cruz, CA 95064
email: mgaskell@ucsc.edu

Abstract. We show that, contrary to simple predictions, most AGNs show at best only a small increase of lags with increasing wavelength in the J, H, K, and L bands. We suggest that a possible cause of this near simultaneity from the near-IR to the mid-IR is that the hot dust is in a hollow bi-conical outflow of which we preferentially see the near side. In the proposed model sublimation or re-creation of dust (with some delay relative luminosity variations) along our line of sight in the hollow cone as the flux varies could be a factor in explaining the AGN changing-look phenomenon (CL). Variations in the dust obscuration can help explain changes in relationship of Hβ time delay on Luv variability. The relative wavelength independence of IR lags simplifies the use of IR lags for estimating cosmological parameters.

Keywords. galaxies: active - galaxies: nuclei - galaxies: Seyfert - infrared: optical and IR variability, time delay, dust, cosmology

1. Introduction

Over half a century has passed since the discovery of optical and near-IR variability of AGNs (Fitch et al. 1967; Pacholczyk 1971; Penston et al. 1971). This was soon followed by the discovery of time delays between optical (U-band) and near-IR variations (Penston et al. 1971, 1974; Lebofsky & Rieke 1980; Clavel et al. 1989; Glass 1992; Baribaud et al. 1992; Oknyanskii 1993). The bolometric luminosity of AGNs is dominated by the "big blue bump", the thermal emission from the inner regions of the accretion disc. The IR radiation of AGNs is mostly thermal emission of dust heated by UV radiation from the inner regions. The time delay is interpreted as a consequence of the light-travel time due to the different locations and sizes of the emitting regions (Barvainis 1987). The hot dust is located away from the axis of rotation (Keel 1980) in what Antonucci (1984) called a "torus". Although the shape and structure of this region of hot dust is uncertain, it is commonly depicted in cartoons as being like a doughnut.

Spatially-resolved IR observations of a number of AGNs show that the dust clouds predominantly emitting in the mid-IR to far-IR range are not concentrated in the plane of the galaxy or of the accretion disc, as might be expected, but in the polar direction (Braatz et al. 1993; Cameron et al. 1993; Bock et al. 2000; Hönig et al. 2012). Spatial resolution for IR emission regions is only possible for the nearest AGNs and mostly in the mid- and far-IR regions. The most common method for studying the unresolved structure of the emission from warm dust emitting in the mid- to near-IR is reverberation mapping using broad-band IR flux data compared with optical/UV fluxes. Such studies have now been carried out for several dozen AGNs. These reveal a square root dependence of the IR time delay on the luminosity (L_{uv}) for a number of AGNs (Oknyanskij 1999; Oknyanskij & Horne 2001; Koshida et al. 2014; Koshida et al. 2017; Minezaki et al. 2019). This potentially gives us a direct way for estimating cosmological parameters (Oknyanskij 1999; Yoshii et al. 2014). Such a luminosity–time delay relationship was predicted theoretically by Barvainis (1987) from radiative equilibrium of the dust. The IR time delays can correspond to the sublimation radius which corresponds to the maximal dust temperature – approximately 17002000 K for graphite grains (Huffman 1977; Salpeter 1977; Baskin & Laor 2018).

In an individual object (for example, NGC 4151) the time delay varies with the mean L_{UV} level. This is because of the destruction and replenishment of dust as the high-energy flux varies. These variations of the IR time delay indicate a lag in changes in the size of the hot dust region of about a few years behind L_{UV} variations (Oknyanskij et al. 2008; Kishimoto et al. 2013). Despite significant growth in theoretical and observational studies of AGNs in the IR, our knowledge of the dust, its origin, kinematics, and detailed morphology remains very incomplete (e.g., Oknyansky et al. 2015; Czerny et al. 2023).

2. Observed IR time lags

Observed IR time delays in different bands have been collected and discussed by Oknyansky et al. (2015); Oknyansky et al. (2019); Koshida et al. (2014); Koshida et al. (2017); Minezaki et al. (2019) and elsewhere. From these results, contrary to simple predictions, we find that most AGNs show at best only a small increase of lags in the J, H, K, and L bands with increasing wavelength. Only for 2 objects, GQ Comae and NGC 4151, was a difference of about a factor of three between lags in the L and K bands observed. For NGC 4151 such a big difference was observed only during the unusually long, very high state which was the only such one during more than a century of monitoring (Oknyanskij et al. 2013). The probability of observing an object exactly when it is in its highest state in a century is not large. Statistically it is much more likely that one observes an object in state which is below a very rare high state. Perhaps this is the reason why we see similar IR lags at different wavelengths for most objects. The relative wavelength independence of IR lags simplifies the use of IR lags for estimating cosmological parameters (Oknyanskij 1999) since it reduces the redshift dependence of the time-lag.

3. A hollow, bi-conical, outflow model

If, as has been widely assumed, the dust clouds are located in the equatorial direction in a flattened, axially-symmetric region (i.e., a "torus"), and if we are viewing the AGN from a "type-1" viewing position not too far off-axis, then the IR lag, τ, gives the effective distance, $c\tau$, of the emitting dust clouds from the central source. In this widely-assumed model, similar lags at different IR wavelengths must be interpreted as implying the existence of dust with different temperatures at approximately the same radii. In Oknyansky et al. (2015) we propose an alternative explanation, that *the dust is in a hollow bi-conical outflow.* (see details there, Oknyansky et al. 2015, 2018).

In our model the relatively dust-free region would have to have large half-opening angle, α (see Fig. 1). The structure of IR emission region is axisymmetric due to axial symmetry of the UV radiation given by $F_{uv}(\theta) \sim [cos(\theta)(1 + 2cos(\theta))/3]^{1/2}$ (see, e.g., Netzer 2015) and also due to dynamic reasons (see, e.g., Wada 2015; Kokubo & Minezaki 2020; Czerny et al. 2023) the center of the cones (O) is below the equatorial plane. Such

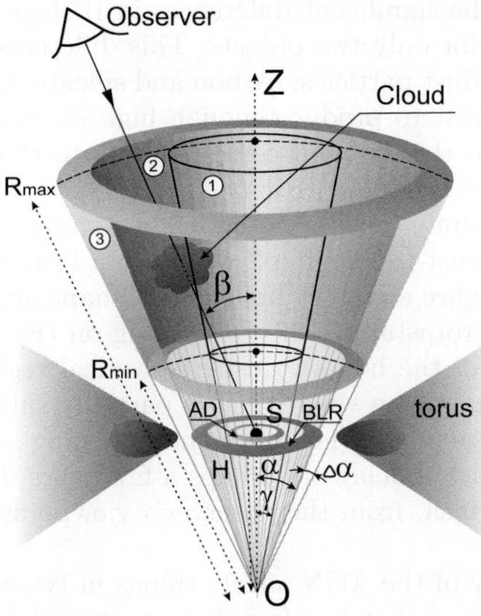

Figure 1. An idealized schematic showing the near side of the proposed geometry of a hollow bi-conical dust distribution. (The extension of the cones below the equatorial plane is explained in the text.) In this idealized model, the observer is viewing the system at an angle β to the axis of symmetry. The three zones of dust sublimation or recreation discussed in the text are indicated. Region number one (half-opening angle $< \gamma$) indicates where the dust is always mostly sublimated and refractory elements are mostly in the gas phase. Region number three (half-opening angle $> \alpha$) corresponds to where the dust will always survive. The half-opening angles γ and α defining the boundaries of the intermediate region (number two) depend on the history of flux variations (see text). In reality, the dust will be in clumps. One such cloud is indicated for illustration purposes and can help explanation the CL events. This gives partial obscuration of the BLR and the accretion disc (AD).

geometry can be connected with suspected local origin of the dust in the accretion disk, and also helps with the need to have as much as possible of UV radiation hitting the dust region. In Oknyansky et al. (2015) we showed that IR lags are similar at different radii in the conical outflow and that this explains the relative wavelength independence of the IR time delay. In Fig 1 we show just the near side of the proposed hollow bi-conical dust distribution. The corresponding cone on the far side (not shown in Fig. 1) gives a much broader response function. Also, the accretion disc (AD) in the equatorial plane is optically thick and the inner parts of the far side of the bi-cone will be hidden behind it.

We consider here what happens in this model when an AGN goes through high and low states. We consider three different general regions of dust sublimation or recreation:

(1) A fully-sublimated dust region within half-opening angle γ. The dust is sublimated here (matter is in the gas phase) even in low states. The dust cannot recover quickly in this region.

(2) A region where the dust is partially sublimated. In this region the dust can survive in some dense parts of clouds where grains are not sublimated to the gas phase. The dust in this region can be recover relatively quickly over a few years when an AGN returns to a low state.

(3) A non-sublimated dust region with a half-opening angle $> \alpha$ located further out than the sublimation zone corresponding to some past very bright state.

Based on what has been observed for NGC 4151, very bright states might happen once per decade or even only once in centuries. If we are lucky and see an object in such an outburst, then we see the sublimation process in this third region. In these very rarely observed events there will be significant differences in IR lags for different wavelengths. As mentioned, we see this for only two objects. This difference in lag can be connected also with different types of dust particles: carbon and silicate. In the bright state, the UV flux can heat the silicate dust to produce enough high re-processing in mid-IR. In this case we see a time delay in the K band corresponding to the *carbon* dust sublimation region but the time delay in L band corresponds to the *silicate* dust sublimation radius.

The geometry will certainly not be spherically symmetric. The mostly-dust-free region 1 and the partially-dust-free region 2 will have a form that can be regarded as a sort of three-dimensional figure eight or "hour-glass" shape such as is seen, for example, in the nebula around the protostar L1527. Depending on the recent luminosity history, the half-opening angle, α, of the hollow, dust-free conical region (region 1), as well as, half-opening angle, γ, of the region where dust is partially sublimated (region 2) will be larger or smaller. The change in size (i.e., of the half-opening angles, α and γ) will follow the change in luminosity with a delay of perhaps a few years. In Fig. 1 we also show the broad-line region (BLR) which, from the observer's viewpoint, is partially obscured by dust.

The change in luminosity of the AGN affects things in two ways. Firstly, the physical size of the BLR will change – what is referred to as "breathing" (see Cackett & Horne 2006) – and secondly, the dust obscuration will change, after some delay of a few years. We see the BLR throuph the dust region and the effective size (depending on dust obscuration) of the BLR has to be following the UV variability with a few years delay. Similar predictions have been made previously (see, e.g., Laor 2004; Gaskell & Harrington 2018) and have been reported for NGC 4151 (Chen et al. 2023). The recovery of dust in some clouds in the region (2) can be much faster than the dynamical time and this can help explain CL events in AGNs.

4. Conclusion

The lags of IR variability behind UV variability do not show the increase of lag with increase of wavelength predicted by simple models. We have proposed a solution in which the dust is in a bi-conical outflow. This not only explains the relative wavelength independence of the IR lags, but the model can be extended to help explain aspects of the AGN "changing-look" phenomenon and the time dependence of the Balmer line lags from L_{uv} variations.

Acknowledgements

We thank A. Cherepashchuk for supporting our research and observations. We are grateful to D. Chelouche, A. Laor, and H. Netzer and for useful discussions. We are thankful to graphic designer Natalia Sinugina for producing Figure 1. This research has been partly supported by Israeli Science Foundation grant no. 2398/19.

References

Antonucci, R. R. J. 1984, ApJ, 278, 499
Baribaud, T., Alloin, D., Glass, I., & Pelat, D. 1992, A&A, 256, 375
Barvainis, R. 1987, ApJ, 320, 537
Baskin, A., & Laor, A. 2018, MNRAS, 474, 1970
Bock, J. J., Neugebauer, G., Matthews, K., et al. 2000, AJ, 120, 2904
Braatz, J. A., Wilson, A. S., Gezari, D. Y., Varosi, F., & Beichman, C. A. 1993, ApJ, 409, L5
Cackett, E. M., & Horne, K. 2006, MNRAS, 365, 1180
Cameron, M., Storey, J. W. V., Rotaciuc, V., et al. 1993, ApJ, 419, 136
Chen, Y.-J., Bao, D.-W., Zhai, S., et al. 2023, MNRAS, 520, 1807
Clavel, J., Wamsteker, W., & Glass, I. S. 1989, ApJ, 337, 236

Czerny, B., Zajaček, M., Naddaf, M.-H., et al. 2023, European Physical Journal D, 77, 56
Fitch, W. S., Pacholczyk, A. G., & Weymann, R. J. 1967, ApJ, 150, L67
Gaskell, C. M., & Harrington, P. Z. 2018, MNRAS, 478, 1660
Glass, I. S. 1992, MNRAS, 256, 23P
Hönig, S. F., Kishimoto, M., Antonucci, R., et al. 2012, ApJ, 755, 149
Huffman, D. R. 1977, Advances in Physics, 26, 129
Keel, W. C. 1980, AJ, 85, 198
Kishimoto, M., Hönig, S. F., Antonucci, R., et al. 2013, ApJ, 775, L36
Kokubo, M., & Minezaki, T. 2020, MNRAS, 491, 4615
Koshida, S., Minezaki, T., Yoshii, Y., et al. 2014, ApJ, 788, 159
Koshida, S., Yoshii, Y., Kobayashi, Y., et al. 2017, ApJ, 842, L13
Laor, A. 2004, in Astronomical Society of the Pacific Conference Series, Vol. 311, AGN Physics with the Sloan Digital Sky Survey, ed. G. T. Richards & P. B. Hall, 169
Lebofsky, M. J., & Rieke, G. H. 1980, Nature, 284, 410
Minezaki, T., Yoshii, Y., Kobayashi, Y., et al. 2019, ApJ, 886, 150
Netzer, H. 2015, ARA&A, 53, 365
Oknyanskii, V. L. 1993, Astronomy Letters, 19, 416
Oknyanskij, V. L. 1999, Odessa Astronomical Publications, 12, 99
Oknyanskij, V. L., & Horne, K. 2001, in Astronomical Society of the Pacific Conference Series, Vol. 224, Probing the Physics of Active Galactic Nuclei, ed. B. M. Peterson, R. W. Pogge, & R. S. Polidan, 149
Oknyanskij, V. L., Lyuty, V. M., Taranova, O. G., Koptelova, E. A., & Shenavrin, V. I. 2008, Odessa Astronomical Publications, 21, 79
Oknyanskij, V. L., Metlova, N. V., Artamonov, B. P., Lyuty, A. V., & Lyuty, V. M. 2013, Odessa Astronomical Publications, 26, 212
Oknyansky, V., Lipunov, V. M., Shatsky, N. I., et al. 2018, The Astronomer's Telegram, 11703
Oknyansky, V. L., Gaskell, C. M., & Shimanovskaya, E. V. 2015, Odessa Astronomical Publications, 28, 175
Oknyansky, V. L., Shenavrin, V. I., Metlova, N. V., & Gaskell, C. M. 2019, Astronomy Letters, 45, 197
Pacholczyk, A. G. 1971, ApJ, 163, 449
Penston, M. V., Penston, M. J., Neugebauer, G., et al. 1971, MNRAS, 153, 29
Penston, M. V., Penston, M. J., Selmes, R. A., Becklin, E. E., & Neugebauer, G. 1974, MNRAS, 169, 357
Salpeter, E. E. 1977, ARA&A, 15, 267
Wada, K. 2015, ApJ, 812, 82
Yoshii, Y., Kobayashi, Y., Minezaki, T., Koshida, S., & Peterson, B. A. 2014, ApJ, 784, L11

Black Hole Winds at All Scales
Proceedings IAU Symposium No. 378, 2024
G. Bruni, M. Diaz Trigo, S. Laha & K. Fukumura, eds.
doi:10.1017/S1743921324000334

Dynamics of M87 jet

Maxim Lyutikov

Department of Physics and Astronomy, Purdue University, 525 Northwestern Avenue, West Lafayette, IN 47907-2036
email: gabriele.bruni@inaf.it

Abstract. Flows originating from black hole magnetospheres via Blandford-Znajek (BZ) process start highly relativistically, with very large Lorentz factors $\gamma_0 \gg 1$, imprinted into the flow during pair production within the gaps. As a result, BZ-driven outflows would produce spine-brightened images, contrary to observations of the edge-brightened jet in M87. We conclude that M87 jet is not BZ-driven.

Keywords. galaxies: active, jets, magnetic fields

1. Introduction

Observations of the inner part of the jet in M87 down to just 7 Schwarzschild radii shows limb brightened collimated jet - the jet accelerating smoothly, with a parabolic profile Nakamura & Asada (2013); Kim et al. (2018); Blandford et al. (2019), Fig. 1. This implies that a flow is only mildly relativistic. For example, estimates of the viewing angle $15^\circ - 30^\circ$ degrees Bicknell & Begelman (1996) imply bulk Lorentz factors $\sim 3 - 2$.

The Blandford & Znajek (1977) mechanism of AGN jet production involves pair production within the rotating black hole magnetosphere ensuing MHD acceleration. We point out (Lyutikov and Haitham, submitted) the dominant importance of plasma injection effects for relativistic winds from black holes. In the highly magnetized case (bead-on-wire) the motion of particles involves electromagnetic drift and motion along the spiral magnetic field. It turns out that a particle launched with large Lorentz factor along the rotating magnetic spiral (in a bead-on-wire approximation) moves nearly radially, Fig. 2.

Since the observed pattern is dominated by Lorentz boost, to calculate the images we employ the following procedure:

• Given the prescribed shape if field lines (e.g. parabolic) and for a given parallel Lorentz factor we calculate local Doppler factor δ. In fact, since toroidal velocity is small, one can use just the shape of the flux surface to find the local direction of the flow.

• We scale local density (somewhat arbitrary) as $1/r^2$, total spherical distance to the black hole.

• local emissivity is parameterized as

$$j \propto \frac{\delta^3}{r^2} \tag{1.1}$$

• emissivity is integrated along the line of sight.

Our results demonstrate a consistent picture: relativistic flows emanating from within the magnetosphere are expected to produce spine-brightened images, Figs 2 and 4.

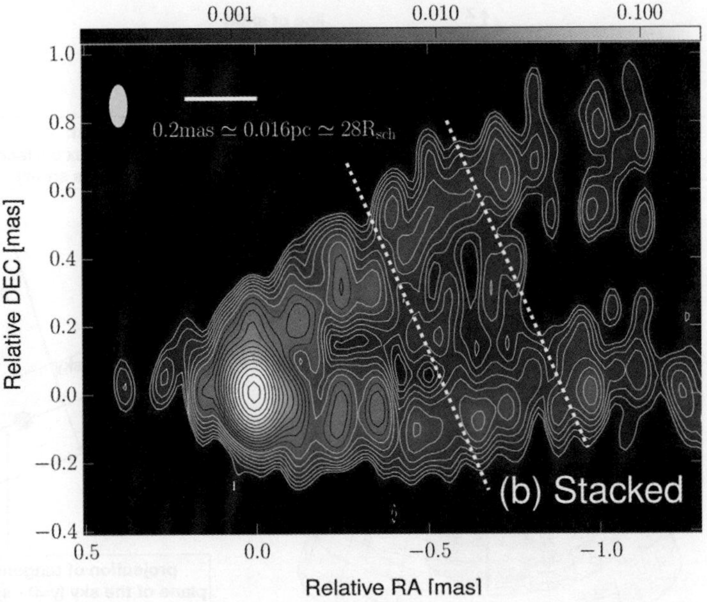

Figure 1. Central part of jet in M87, showing edge-brightened structure Kim et al. (2018).

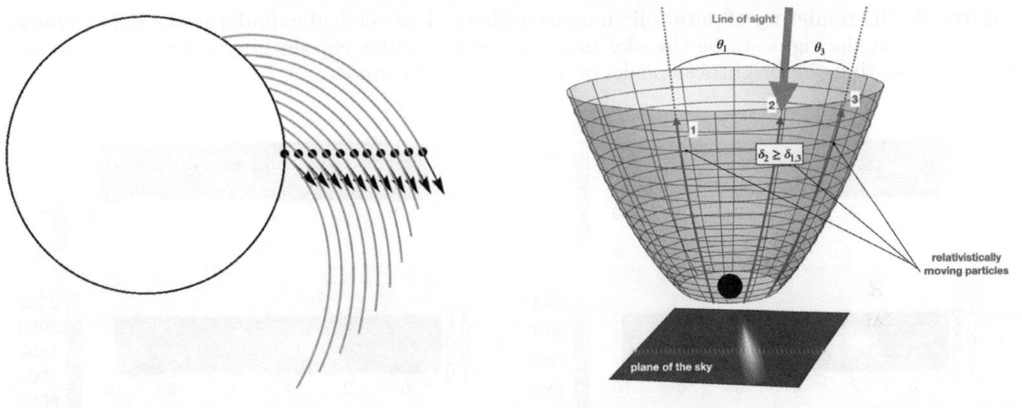

Figure 2. Left Panel: Particles' trajectories for Lorentz factors $\gamma_0 10$. The circle is the light cylinder, orange curves indicate the same magnetic field line at consecutive movements, arrows are directions of particle velocity at each point, and black dots are the location of the particle. These calculations illustrate that for $\gamma_0 \gg 1$ the trajectory is nearly radial. Graphic explanation why BH-produced jets are expected to be spine-brightened. Shown are flux surfaces and three different particles' trajectories. Emission produced by particle 2 has the smallest angle with respect to the line of sight, largest Doppler factor, and would result in brightest emission pattern.

2. Conclusion

2.1. *Morphology of M87 jet and the BZ mechanism*

Our results show a universal property of the resolved Blandford & Znajek (1977) flow with large parallel (to the local magnetic field) momentum of emitting particles: all images are spine-brightened, not edge-brightened as observed. No special prescription for emissivity as function of the flux function parameter r_0 can change that: they are all spine-brightened. The assumption of parabolical flux surfaces is, naturally, an analytic approximation, yet the universality of the result - spine-brightened profile - ensures that it will be applicable to more general cases.

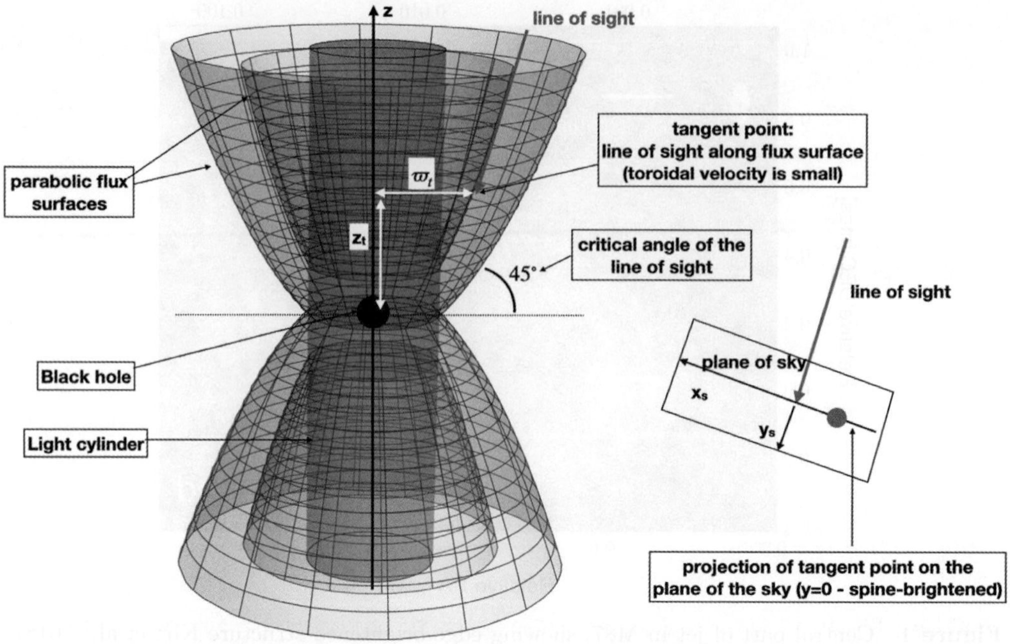

Figure 3. 3D rendering of parabolic magnetosphere. The vertical cylinder is the light cylinder. The insert on the right shows the sky image for a particular chosen line of sight. Due to high Doppler boosting the brightness peaks on the y axis - the spine.

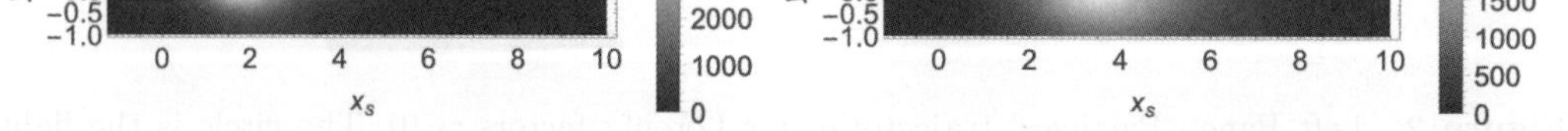

Figure 4. Image maps for emission from different magnetic flux surfaces, observer angle $\theta_{ob} = \pi/12$. The axes correspond to projected distances measured in terms of the light cylinder radius. All images are spine-brightened.

We then conclude that Blandford & Znajek (1977) mechanism is not responsible for the M87 jet, at least in its pure form. †

The Blandford & Znajek (1977) mechanism can still be operational - and, e.g. responsible for the bright core in the images of M87 in case of purely radial outflow at small r, but it does not drive the observed jet. Another possibility is that the sheath is slowed down by the interaction with the disk corona - while the core remains relativistic, with emission beamed away. This would correspond to emission coming only from innermost flux surfaces, top row in Fig. 4 - it is still spine-brightened, but shows only in the small part of the image.

† By *BZ mechanism* we understand generation of collimated relativistic jet. The relativistic $e^{\pm}$ we describe do extract energy from the spin of the black hole.

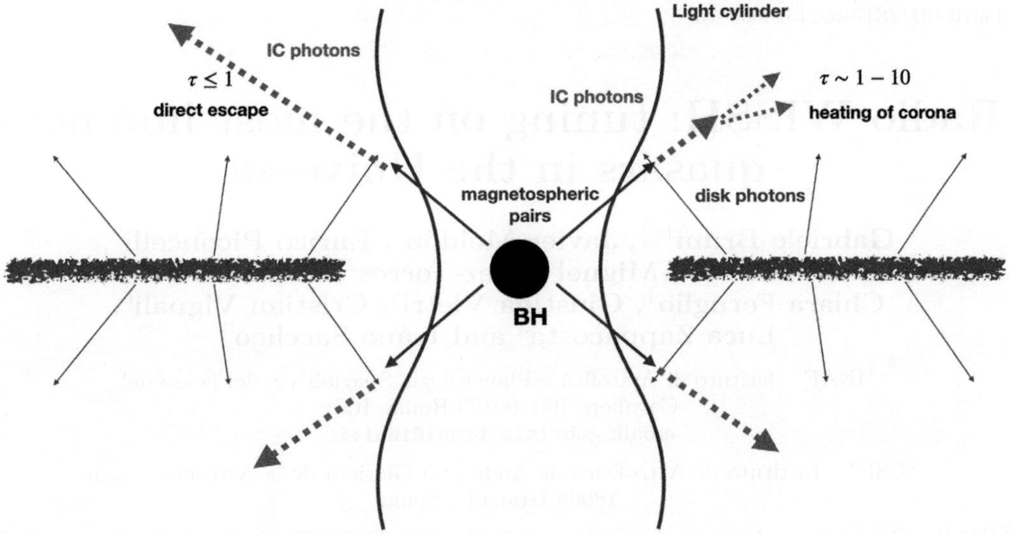

Figure 5. Cartoon of the IC emission by accreting black hole. Particles accelerated with the black hole magnetosphere to $\gamma_0 \gg 1$ IC scatter the soft photons from the disk. The resulting signal is (i) quasi-isotropic; (ii) highly variable due to temporal variations of the gap-produced particle flux.

In contrast disk-produced outflows Blandford & Payne (1982) start non-relativistically. The extended structure observed in M87 is thus inconsistent with the magnetosphere-produced jet, but is consistent with the disk-produce jets.

2.2. *Blazar-like phenomena from non-aligned ANGe*

The observation of rapidly variable very high energy (VHE) gamma-rays from non-aligned active galactic nuclei e.g., from M87 and Cen A, proves challenging for conventional theoretical acceleration and emission models Aharonian et al. (2006); Acciari et al. (2008); Rieger & Aharonian (2008). Even for the jet-aligned Blazers variations down to a few minutes time scale, smaller than the expected size of the black hole Aharonian et al. (2005), are difficult to explain.

The present model provides a possible answer, Fig. 5: (i) particles are streaming relativistically, radially and quasi-isotropically *right in the vicinity of the black hole*. If the flow varies on time scale $\sim 1/\Omega_{BH}$, then the typical time scale of variability t_{var} is

$$t_{var} \sim (c/\Omega_{BH}) \times 1/\gamma_0^2 = a_{Kerr}^{-1} \times R_{BH}/c \times 1/\gamma_0^2 \tag{2.1}$$

where a_{Kerr} is the Kerr spin parameter and R_{BH} is the horizon size of the black hole. Since the expected Lorentz factor γ_0 can be in the thousands, the variability time scale cane be very short.

References

Acciari, et al. 2008, *ApJ*, 679, 397
Aharonian, et al. 2005, *A&A*, 442, 895
Aharonian, et al. 2006, Science, 314, 1424
Bicknell, G. V., & Begelman, M. C. 1996, *ApJ*, 467, 597
Blandford, et al. 2019, *ARAA*, 57, 467
Blandford, R. D., & Payne, D. G. 1982, *MNRAS*, 199, 883
Blandford, R. D., & Znajek, R. L. 1977, *MNRAS*, 179, 433
Kim, 2018, *A&A*, 616, A188
Nakamura, M., & Asada, K. 2013, *ApJ*, 775, 118
Rieger, F. M., & Aharonian, F. A. 2008, *A&A*, 479, L5

Black Hole Winds at All Scales
Proceedings IAU Symposium No. 378, 2024
G. Bruni, M. Diaz Trigo, S. Laha & K. Fukumura, eds.
doi:10.1017/S1743921323003575

Radio WISSH: tuning on the most luminous quasars in the Universe

Gabriele Bruni[1], Javier Moldón[2], Enrico Piconcelli[3], Francesca Panessa[1], Miguel Pérez-Torres[2], Manuela Bischetti[4], Chiara Feruglio[4], Giustina Vietri[5], Cristian Vignali[6], Luca Zappacosta[3] and Ivano Saccheo[3]

[1]INAF – Istituto di Astrofica e Planetologia Spaziali via del Fosso del Cavaliere 100, 00133 Roma, Italy
email: gabriele.bruni@inaf.it

[2]CSIC – Instituto de Astrofísica de Andalucía Glorieta de la Astronomía s/n, 18008 Granada, Spain

[3]INAF – Osservatorio Astronomico di Roma via Frascati 33, 00078 Monte Porzio Catone, Italy

[4]INAF – Osservatorio Astronomico di Trieste via Tiepolo 11, 34143 Trieste, Italy

[5]INAF – Istituto di Astrofisica Spaziale e Fisica Cosmica di Milano via A. Corti 12, 20133 Milano, Italy

[6]Dipartimento di Fisica e Astronomia "Augusto Righi", Università degli Studi di Bologna via Gobetti 93/2, 40129 Bologna, Italy

Abstract. In the past years, the results obtained by the WISSH quasar project provided a novel general picture on the distinctive multi-band properties of hyper-luminous ($L_{bol} > 10^{47}$ erg/s) quasars at high redshift (z $\sim$ 2-4), unveiling interesting relations among active galactic nuclei, winds and interstellar medium, in these powerful sources at cosmic noon. Since 2022, we are performing a systematic and statistically-significant VLA study of the radio properties of WISSH. We carried out high-resolution VLA observations aiming at: 1) identifying young radio source from the broad-band spectral shape of these objects; 2) sample an unexplored high redshift/high luminosity regime, tracking possible evolutionary effects on the radio-loud/radio-quiet dichotomy; 3) quantifying orientation effects on the observed winds/outflows properties.

Keywords. quasar, AGN, outflows, super-massive black hole, ISM: jets and outflows

1. Introduction

Hyper-luminous QSOs (HyLQSOs, i.e. $L_{bol} > 10^{47}$ erg/s), powered by the most massive, highly-accreting supermassive black holes (SMBHs, i.e. $M_{BH} > 10^9\ M_\odot$), are ideal targets to probe the assembly of giant galaxies (and, likely, the cradles of proto-clusters). Following the nowadays consensus view on SMBH-host galaxy co-evolution, the huge amount of energy released by highly-accreting SMBHs in HyLQSOs is able to strongly affect the evolution of the host galaxy by heating and expelling the interstellar medium (ISM) (the so-called "AGN feedback" mechanism, see e.g. Fabian 2012, Morganti 2017 for a review). The systematic study of HyLQSOs has faced significant challenges due to their low number density and low fluxes resulting from their distance. A significant improvement in our understanding of the properties of the accreting SMBH ($M_{\rm BH}$, $\lambda_{\rm Edd}$), the nuclear thermal and non-thermal emission components, multiphase winds and multiphase ISM (and their interplay) in HyLQSOs can be only achieved investigating all these aspects in a large sample of HyLQSOs. Namely, one that is markedly different from traditional observing programs in a specific frequency band, and focused on sparse sources to study a particular aspect of the HyLQSO phenomenon. This highlights the necessity of building large samples of HyLQSOs with extensive multi-band coverage from radio to X-rays. The additional information coming from the radio band can provide fundamental inputs on the presence of jets, their interplay with winds, and in general on the presence of a possible young radio phase at cosmic noon, and the long-standing questions about the radio-loud/radio-quiet dichotomy across cosmic epochs.

1.1. *Probing the brightest end of the AGN luminosity function with WISSH*

The WISSH quasar project can be regarded as a multi-band effort in the study of HyLQSOs, as demonstrated by the number of publications since 2017 dealing with their the central engine, the outflows/feedback and host galaxy properties, e.g. Bischetti et al. (2017); Duras et al. (2017); Martocchia et al. (2017); Bischetti et al. (2018); Vietri et al. (2018); Bruni et al. (2019); Travascio et al. (2020); Zappacosta et al. (2020); Bischetti et al. (2021); Vietri et al. (2022). The aim of the WISSH project is to establish a reference sample of HyLQSOs at cosmic noon to investigate their nuclear properties and the AGN feedback mechanism on a sound statistical basis. The sample consists of 85 broad-line Type 1, radio-quiet AGN at $z \sim 2$–4.5 from SDSS-DR7 and selected by the WISE All Sky Survey with flux $F_{22\mu m} > 3$ mJy (see Saccheo et al. 2023 and references therein). Accordingly, the WISSH quasars result to be among the most luminous AGN known in the Universe with $L_{\rm Bol} > 2 \times 10^{47}$ erg s^{-1}. In this proceeding, we briefly summarize the first results of our radio campaign on WISSH QSOs, aiming at characterizing the radio emission in objects at cosmic noon.

2. A radio characterization of the WISSH sample

During 2022, we realized a deep, high-resolution VLA survey of the WISSH sample in A configuration in the 2-8 GHz range. We covered ~90% of the sample at 2-4 GHz, while ~75% at 4-8 GHz, probing physical scales between 2 and 5 kpc. Our strategy was to reach a sensitivity threshold of ~50 μJy, well below past or current radio surveys. Indeed, a first radio approach to the WISSH sample carried out by Bruni et al. (2019) showed that, cross-correlating with the FIRST survey at 1.4 GHz (Becker et al. 1995), only ~20% of the objects shows a detection at a ~500 μJy sensitivity threshold, and a compact morphology (<40 kpc at the median redshift of the sample). The recent release of the first epoch of the VLASS survey at 3 GHz - at a sensitivity similar to the FIRST one (RMS~120 μJy/beam) - allowed us to confirm this detection rate. The estimated radio loudness ($R = f_{6cm}/f_{4400\mathring{A}}$) is lower than 10 for all except two objects, for which R = 47 and 290. Given these premises, going deeper in terms of physical scale and sensitivity appeared to be key to unveil the radio properties of the sample. The main goals of our VLA campaign are the following:

- Quantify the fraction of young radio sources: by compiling the $L_{1.4\rm GHz}$ vs linear size (LS) diagram, we will be able to compare with the fraction of young radio sources found for heavily obscured quasars (Patil et al. 2020), investigating how the different evolutionary stages influence the radio phase. The collected VLA data can be complemented with LOFAR measurements from the LoTSS DR2 survey (Shimwell et al. 2022) - where, among the 34 sources in the footprint, 32 were detected - allowing us to extend the frequency coverage down to 0.15 GHz. The overall radio spectrum will allow us to quantify the fraction of peaked sources in the 0.15-12 GHz range, possibly confirming the fraction of young radio sources estimated from the $L_{1.4\rm GHz}$ vs LS diagram.

• Probe an unexplored high redshift/high luminosity regime of active galaxies: at the median redshift of WISSH ($z = 3.33$), and with an RMS of ~10 μJy/beam at 3 GHz, it is possible to probe radio powers down to $\sim 7 \times 10^{23}$ W/Hz at 3-sigma significance. Thanks to the availability of optical and X-ray (Chandra program ongoing) luminosities, a distribution of radio-loudness can be obtained and compared to those of other surveys in order to test a possible evolution scenario of the radio loudness (Ballo et al. 2012). This can provide clues on long-standing questions about the radio-loud/radio-quiet dichotomy.

• Test orientation effects on the observed winds/outflows properties: in quasars, the spectral index of the optically-thin part of the radio jet spectrum can be used as an indicator of the jet orientation (Orr & Browne 1982), suggesting a near-to-polar line of sight for values > -0.5, while an equatorial one for values < -0.5 ($S_\nu \propto \nu^\alpha$). This information can be compared with the outflows orientation estimates from Vietri et al. (2018), and with the presence of nuclear winds (BAL) from Bruni et al. (2019).

2.1. *First results: detection rates and morphologies*

At an RMS of ~10 μJy/beam, about 80% of the observed sample was detected at 3 GHz - reaching a redshift of 4.3 - implying an estimated radio power $> 10^{23}$ W/Hz (classical threshold between radio-quiet and radio-loud AGN in the local Universe, Condon 1992). This suggests that, at cosmic noon, most of the hyperluminous AGN like the WISSH ones could host jets. The implication of this result are wide-reaching, from the evolutionary effects on the radio-quiet/radio-loud dichotomy, to the contribution of jets in the QSOs feedback budget at large redshifts, and jets launching at this luminosity regime. The very fact that most of sources lie below the VLASS detection threshold (see Fig. 1, left panel) highlights how important deep observations are to perform population study at this high redshift, allowing to reach the completeness needed to draw conclusions on the radio phase evolution in AGN.

Three objects showed a resolved morphology at 3 GHz, and more could arise from higher frequency observations. They show a symmetric morphology centered on the optical position of the host, with a projected linear extension of about 30 kpc (see Fig. 1, right panel). This could suggest that these sources are radio galaxies at cosmic noons, but more analysis, including spectral index estimates, is necessary to claim this.

Observations are still ongoing, and will be concluded at the end of the current VLA semester, completing the multi-frequency radio view on the WISSH sample.

2.2. *The radio phase across cosmic epochs*

Recently, Patil et al. (2020) performed a VLA study of heavily obscured quasars. The sample was selected in the ultra-luminous regime ($L_{bol} \sim 10^{11.7-14.2} L_\odot$) at $z \sim 0.4 - 3$, and extremely red in the WISE mid-infrared/optical band, along with a detection of bright, unresolved radio emission from the NVSS. Thanks to high-resolution VLA observations at 10 GHz, they found radio luminosities and linear extents similar to young radio sources (Gigahertz Peaked Spectrum, GPS, and Compact Steep Spectrum, CSS, sources). In a subsequent paper (Patil et al. 2022), they built the radio spectra by adding data from surveys, and confirmed the presence of a high fraction of young radio source.

Although both the Patil et al. (2020) and WISSH quasar samples are selected to allow the direct observation of AGN feedback in action, they are complementary in terms of quasar evolutionary stages. Indeed, according to Hopkins et al. (2008), the obscured quasars in Patil et al. (2020) represent the initial heavily dust-enshrouded phase associated with rapid SMBH growth and star formation triggered by multiple galaxy encounters, while optically-bright objects like WISSH ones are undergoing the

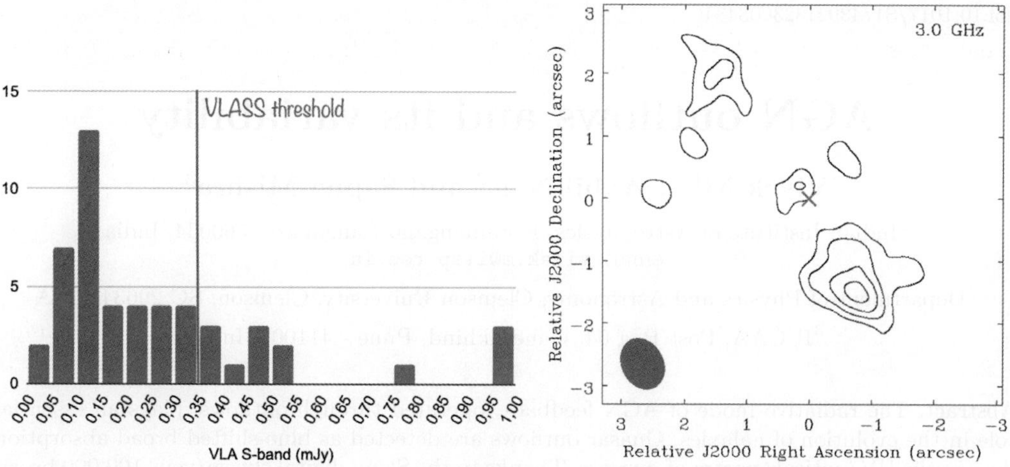

Figure 1. *Left panel* – Radio flux densities at 3 GHz for the WISSH sample (truncated at 1 mJy). Most of objects lie below the VLASS detection threshold. *Right panel* – A resolved source from the sample. The cross indicates the optical position of the host, the FWHM is reported in the lower-left corner. The projected linear size of the source is ∼40 kpc.

"blow-out" phase, which is characterized by powerful QSO-driven outflows blowing away the nuclear dust cocoon and part of the cold gas reservoir in the host galaxy.

The same kind of study performed by Patil et al. (2020) and Patil et al. (2022), once realized on the WISSH sample of hyper-luminous broad-line quasars, will not only provide unprecedented information on the radio phase at cosmic noon, but also shed light on the possible link between the launching mechanism of nuclear winds and radio jets.

References

Ballo, L., Heras, F. J. H., Barcons, X., et al. 2012, *A&A*, 545, A66
Becker, R. H., White, R. L., & Helfand, D. J. 1995, *ApJ*, 450, 559
Bischetti, M., Piconcelli, E., Vietri, G., et al. 2017, *A&A*, 598, A122
Bischetti, M., Piconcelli, E., Feruglio, C., et al. 2018, *A&A*, 617, A82
Bischetti, M., Feruglio, C., Piconcelli, E., et al. 2021, *A&A*, 645, A33
Bruni, G., Piconcelli, E., Misawa, T., et al. 2019, *A&A*, 630, A111
Condon, J. J. 1992, *ARAA*, 30, 575
Duras, F., Bongiorno, A., Piconcelli, E., et al. 2017, *A&A*, 604, A67
Fabian, A. C. 2012, *ARAA*, 50, 455
Hopkins, P. F., Cox, T. J., Kereš, D., et al. 2008, *ApJS*, 175, 390
Martocchia, S., Piconcelli, E., Zappacosta, L., et al. 2017, *A&A*, 608, A51
Morganti, R. 2017, *Frontiers in Astronomy and Space Sciences*, 4, 42
Orr, M. J. L. & Browne, I. W. A. 1982, *MNRAS*, 200, 1067
Patil, P., Nyland, K., Whittle, M., et al. 2020, *ApJ*, 896, 18
Patil, P., Whittle, M., Nyland, K., et al. 2022, *ApJ*, 934, 26
Saccheo, I., Bongiorno, A., Piconcelli, E., et al. 2023, *A&A*, 671, A34
Shimwell, T. W., Hardcastle, M. J., Tasse, C., et al. 2022, *A&A*, 659, A1
Travascio, A., Zappacosta, L., Cantalupo, S., et al. 2020, *A&A*, 635, A157
Vietri, G., Piconcelli, E., Bischetti, M., et al. 2018, *A&A*, 617, A81
Vietri, G., Misawa, T., Piconcelli, E., et al. 2022, *A&A*, 668, A87
Zappacosta, L., Piconcelli, E., Giustini, M., et al. 2020, *A&A*, 635, L5

Black Hole Winds at All Scales
Proceedings IAU Symposium No. 378, 2024
G. Bruni, M. Diaz Trigo, S. Laha & K. Fukumura, eds.
doi:10.1017/S1743921323003484

AGN outflows and its variability

Vivek M[1], Akhil Nair[2] and Sapna Mishra[3]

[1]Indian Institute of Astrophysics, Koramangala, Bangalore - 560034, India
email: vivek.m@iiap.res.in

[2]Department of Physics and Astronomy, Clemson University, Clemson, SC 29634, USA

[3]IUCAA, Post Bag 04, Ganeshkhind, Pune - 411007, India

Abstract. The radiative mode of AGN feedback, operated through outflows, plays an essential role in the evolution of galaxies. Quasar outflows are detected as blue-shifted broad absorption lines in the UV/optical spectra of quasars. Thanks to the Sloan digital sky survey, 100,000 broad absorption line quasars are available now for ensemble statistical studies. This rich dataset has also enabled us to identify some peculiar cases of these sources. By quantifying the BAL fraction in radio-loud BAL quasars, our studies demonstrate a clear trend of increasing BAL fraction as the viewing angle approaches an edge-on orientation, favoring the orientation model of BAL quasars. Also, by contrasting the properties of BAL quasars with appearing and disappearing BAL troughs, our analysis suggests that the extreme variations in BAL troughs are driven by ionization changes.

Keywords. Active Galaxies, Quasar outflows, Broad Absorption Lines

1. Introduction

AGN feedback is a self-regulating process that connects the energy released by an Active Galactic Nucleus (AGN) to the surrounding medium, thereby influencing the evolution of the host galaxy and its central supermassive black hole (SMBH). A compelling piece of evidence supporting AGN feedback is the observed tight correlation between the bulge mass of a galaxy and the mass of its central SMBH (Ferrarese & Merritt (2000), Gebhardt *et al.* (2000)). Simulations have demonstrated that AGN can exert "negative" feedback through mechanisms such as heating up or blowing away the interstellar medium (ISM) (Kauffmann and Haehnelt (2000), Di Matteo, Springel, and Hernquist (2005)). There are two primary modes of AGN feedback: the quasar mode (radiative or wind mode) and the radio mode (kinetic mode), as discussed in the review by Fabian (2012). The quasar mode is prevalent in high-luminosity AGN and involves outflowing winds, while the radio mode occurs in low-luminosity AGN and is associated with radio jets. Compared to radio jets, the wide opening angles in quasar winds, deduced from the detection rates of quasar outflows, enable efficient interaction with the surrounding medium. Consequently, comprehending the radiative mode of AGN feedback, particularly the nature of outflows, holds significant importance in constraining various feedback mechanisms and models of galaxy evolution.

The presence of high-velocity outflows can be inferred from blue-shifted broad absorption lines (BALs) observed in the spectra of 10-20% of quasars (Weymann *et al.* (1991)). BAL quasars are categorized into three sub-classes based on the ionization state of the absorbing gas: high-ionization BALs (Hi-BALs), low-ionization BALs (LoBALs), and iron-LoBALs (FeLoBALs) (Filiz Ak *et al.* (2013), Vivek Srianand & Dawson (2018),

Vivek (2019)). The fraction of observed BAL quasars can be explained either by an orientation model, where the line of sight intersects with the BAL-absorbing clouds, or an evolutionary model, where the quasar goes through a BAL phase for a certain period (Elvis (2000), Farrah *et al.* (2007)).

2. BALQSO fraction

The fraction of BALQSOs provides valuable insights into the orientation versus evolution models of BALQSOs. Defining BALQSOs is an important consideration in this context. Traditionally, BALs are defined as having C IV absorption troughs broader than 2000 km/s and specific outflow velocities. The fraction of "traditional" BALQSOs in optically selected quasar samples is commonly observed to be approximately 10 percent (Weymann *et al.* (1991); Gibson *et al.* (2008)). While this definition is useful for including nuclear outflows, it excludes a significant portion of genuine BALQSOs. To overcome this limitation, the definition of BALQSO was revised by Trump *et al.* (2006) (Trump et al., 2006) to incorporate absorption features characterized by narrower widths and lower outflow velocities. By employing this revised definition, they discovered a significant augmentation in the BALQSO fraction, reaching 26 percent. Studies utilizing infrared or radio selection have also found the intrinsic fraction to be closer to 20 percent (Ganguly & Brotherton (2008), Dai *et al.* (2012), Morabito *et al.* (2014)).

The nature of BAL quasars is still not fully understood, primarily due to the lack of a reliable orientation indicator for AGN. Statistical indicators based on radio, UV, and optical properties have been used. Studies utilizing simulations have indicated a potential relationship between the velocity offsets and line width ratio of high-ionization C IV and low-ionization Mg II lines, suggesting their potential as indicators of viewing angles (Yong et al., 2020). Studies have also investigated emission line properties, such as the Hβ full width at half maximum (FWHM) and O III equivalent width, as potential orientation indicators for low-redshift quasars (Shen and Ho, 2014).

In the standard AGN unification theory, anisotropic radio emission from relativistic jets suggests orientation-dependent properties. Two commonly used radio indicators are the core-to-lobe flux density ratio (R) and the radio-to-optical ratio of the quasar core (R_I). The R parameter depends on core and lobe radio flux densities, while R_I normalizes the core's radio flux by its optical flux. Both indicators show correlations with orientation, but other factors introduce scatter in the correlations.

In our study Nair & Vivek (2022), our objective was to examine the orientation model of BAL quasars by utilizing a sample of sources that were present in both the SDSS Data Release (DR)-16 quasar catalog and the VLA-Faint Images of the Radio Sky at Twenty Centimeters (FIRST) survey. To accomplish this, we utilized radio cut-out images obtained from the FIRST survey and developed a deep learning model based on convolutional neural networks (CNN). This model enabled us to classify the radio morphologies of quasars into categories such as core-only, young jet, single lobe, or triples. Additionally, we further categorized these radio morphologies into core-dominated and lobe-dominated sources.

With our CNN models, we achieved a high precision of over 98% for classifying the various morphological sub-classes of quasars. The average fraction of BALs in the resolved core, core-dominated, and lobe-dominated quasars aligns with the BAL fraction observed in radio and infrared surveys. Furthermore, we analyzed the distribution of BAL quasars based on quasar orientation using radio core-dominance as an indicator. This analysis was also performed for sub-classes of HiBALs, LoBALs, and FeLoBALs.

Fig. 1 shows the BAL fraction as a function of $\log(R_I)$ is examined for different categories of quasars: resolved core quasars (black;dotted), core-dominated quasars

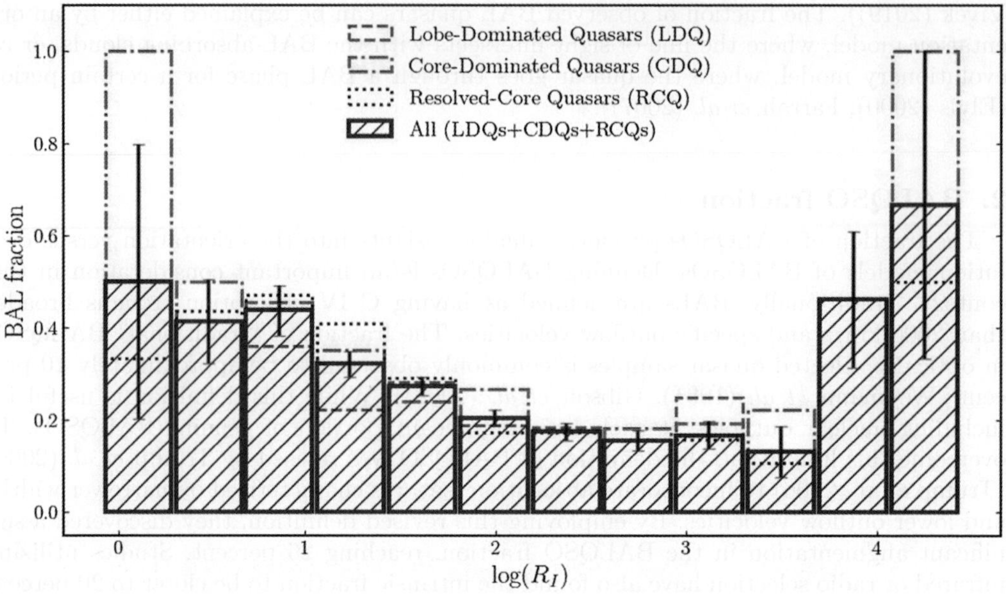

Figure 1. The relationship between BAL fraction and log(R_I) is examined for different categories of quasars: resolved core quasars (black-dotted), core-dominated quasars (red-dot-dashed), lobe-dominated quasars (green-dashed), and a combined sample of resolved core, core-dominated, and lobe-dominated quasars (blue-hatched). Error bars representing 1σ bootstrap errors are included for the combined sample, providing a measure of uncertainty in the BAL fraction.

(red;dot-dashed), lobe-dominated quasars (green;dashed), and a combined sample of resolved core, core-dominated, and lobe-dominated quasars (blue;hatched).

Our findings reveal that all the radio morphological sub-classes and BAL sub-classes exhibit an increased BAL fraction when the jets' orientation angles are closer to the line of sight. This suggests that BAL quasars are more likely to be observed when the viewing angles are in proximity to the equatorial plane of the quasar. However, it is important to note that a pure orientation model alone is insufficient to explain the entirety of the BAL phenomena, and a combination of orientation and evolution is likely the most comprehensive explanation.

3. Variability of BAL troughs

Multiyear variability studies of BAL quasars is crucial in order to gain insights into the location and physical characteristics of the absorbing gas, such as the lifetimes, sizes, and geometries of quasar winds, as well as the underlying mechanisms responsible for these outflows. By examining these variability patterns and their underlying causes, we can gain a deeper understanding of BAL quasars and the physical processes governing their behavior. Various aspects of BAL trough variations, including changes in absorption strength (e.g., equivalent width), the appearance or disappearance of BAL troughs, kinematic shifts in absorption profiles, and alterations in the shape of C IV and Si IV BAL profiles, have been investigated in several cases.

The observed variations in BAL absorption profiles can be attributed to two main factors. Firstly, changes in the covering fraction of the quasar by the absorbing gas can occur as a result of its transverse motions across our line of sight. This has been studied by researchers such as Capellupo *et al.* (2013), Vivek *et al.* (2014). Secondly, variations in the ionizing radiation can lead to changes in the optical depth of the absorbers. This

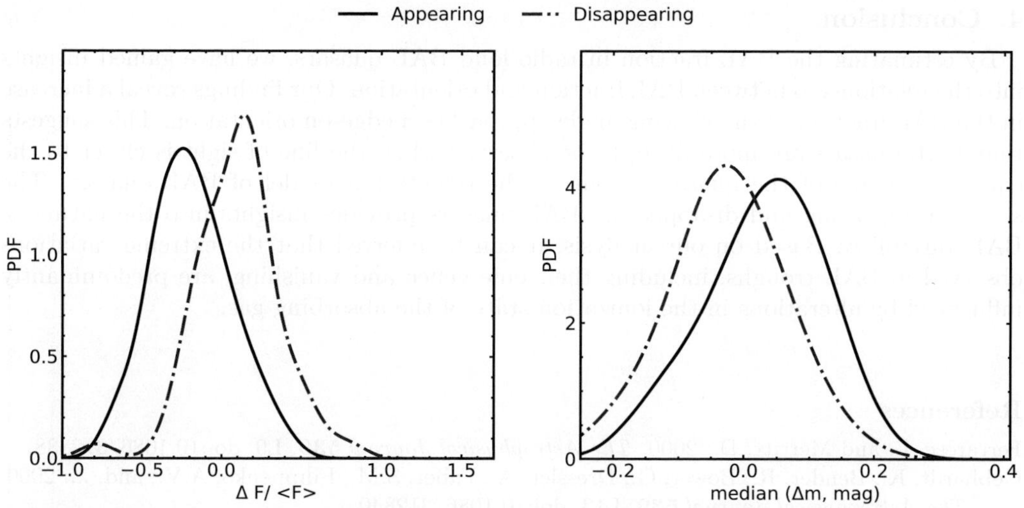

Figure 2. Distribution of (a) change in continuum flux at 1700 Å(b) median of change in V-band magnitudes for Appearing (solid),and Disappearing (dot-dashed) samples.

factor has been explored by Rogerson *et al.* (2018), and Vivek (2019). Additionally, changes in the acceleration profile and/or geometry of the outflow due to alterations in the driving force can also contribute to the observed variability. Studies by Grier *et al.* (2016), and Lu and Lin (2019) have investigated this aspect.

In our recent study, Mishra *et al.* (2021), we presented a new set of 84 broad absorption line (BAL) quasars ($1.7 < z_{em} < 4.4$) that exhibited an appearance of C IV BAL troughs over a span of 0.3-4.8 rest-frame years. To investigate the characteristics of BAL variability, we conducted a comparative analysis using the SDSSDR-7, SDSSDR-12, and SDSSDR-14 quasar catalogues. Our study focused on examining the nature of BAL variability in an appearing BAL quasar sample and contrasting it with a disappearing BAL quasar sample previously studied in the literature. By comparing the two samples, we examined the intrinsic properties, BAL trough characteristics, and continuum parameters of the quasars.

Our findings indicate that the appearing BAL quasars exhibited higher redshift values and shorter probed time-scales in comparison to the disappearing BAL quasars., ensuring that they had comparable distributions of redshift values. We discovered that the appearing BAL quasars were relatively brighter and had shallower and wider BAL troughs compared to the disappearing BAL sample. The distribution of quasar continuum variability parameters exhibited distinct separation between the two samples, indicating that the presence of BAL troughs coincided with a dimming of the continuum, and conversely, the absence of BAL troughs correlated with a brighter continuum. Fig. 2 shows the distribution of change in continuum flux at 1700 Å and the median of change in V-band magnitudes for appearing (solid), and disappearing (dot-dashed) BALQSO samples. The observed spectral index variations in both samples indicated a consistent anti-correlation between the BAL trough and continuum variations, aligning with the well-known trend of "bluer when brighter" in quasars.

We showed that the intrinsic dust model was less likely to be a favorable scenario in explaining BAL appearance/disappearance. Our analysis suggested that the extreme variations of BAL troughs, such as BAL appearance/disappearance, were mainly driven by changes in the ionization conditions of the absorbing gas.

4. Conclusion

By estimating the BAL fraction in radio-loud BAL quasars, we have gained insights into the relationship between BAL fraction and orientation. Our findings reveal a increase in the BAL fraction as the viewing angle approaches a edge-on orientation. This suggests that BAL quasars are more likely to be observed when the line of sight is closer to the equatorial plane of the quasar, favouring the orientation model of BAL quasars. The study of appearing and disappearing BAL quasars provides insights into the nature of BAL variability. Based on our analysis, it can be inferred that the extreme variations observed in BAL troughs, including their emergence and vanishing, are predominantly influenced by alterations in the ionization state of the absorbing gas.

References

Ferrarese, L. and Merritt, D.: 2000, *The Astrophysical Journal* **539**, L9. doi:10.1086/312838.

Gebhardt, K., Bender, R., Bower, G., Dressler, A., Faber, S.M., Filippenko, A.V., and, ...: 2000, *The Astrophysical Journal* **539**, L13. doi:10.1086/312840.

Kauffmann, G. and Haehnelt, M.: 2000, *Monthly Notices of the Royal Astronomical Society* **311**, 576. doi:10.1046/j.1365-8711.2000.03077.x.

Di Matteo, T., Springel, V., and Hernquist, L.: 2005, *Nature* **433**, 604. doi:10.1038/nature03335.

Fabian, A.C.: 2012, *Annual Review of Astronomy and Astrophysics* **50**, 455. doi:10.1146/annurev-astro-081811-125521.

Weymann, R.J., Morris, S.L., Foltz, C.B., and Hewett, P.C.: 1991, *The Astrophysical Journal* **373**, 23. doi:10.1086/170020.

Vivek, M.: 2019, *Monthly Notices of the Royal Astronomical Society* **486**, 2379. doi:10.1093/mnras/stz947.

Vivek, M., Srianand, R., and Dawson, K.S.: 2018, *Monthly Notices of the Royal Astronomical Society* **481**, 5570. doi:10.1093/mnras/sty2712.

Filiz Ak, N., Brandt, W.N., Hall, P.B., Schneider, D.P., Anderson, S.F., Hamann, F., and, ...: 2013, *The Astrophysical Journal* **777**, 168. doi:10.1088/0004-637X/777/2/168.

Elvis, M.: 2000, *The Astrophysical Journal* **545**, 63. doi:10.1086/317778.

Farrah, D., Lacy, M., Priddey, R., Borys, C., and Afonso, J.: 2007, *The Astrophysical Journal* **662**, L59. doi:10.1086/519492.

Gibson, R.R., Brandt, W.N., Schneider, D.P., and Gallagher, S.C.: 2008, *The Astrophysical Journal* **675**, 985. doi:10.1086/527462.

Trump, J.R., Hall, P.B., Reichard, T.A., Richards, G.T., Schneider, D.P., Vanden Berk, D.E., and, ...: 2006, *The Astrophysical Journal Supplement Series* **165**, 1. doi:10.1086/503834.

Ganguly, R. and Brotherton, M.S.: 2008, *The Astrophysical Journal* **672**, 102. doi:10.1086/524106.

Dai, X., Morabito, L.K., Shankar, F., Sivakoff, G.R., and Leighly, K.M.: 2012, *AGN Winds in Charleston* **460**, 120.

Morabito, L.K., Dai, X., Leighly, K.M., Sivakoff, G.R., and Shankar, F.: 2014, *The Astrophysical Journal* **786**, 58. doi:10.1088/0004-637X/786/1/58.

Kimball, A.E., Ivezić, Ž., Wiita, P.J., and Schneider, D.P.: 2011, *The Astronomical Journal* **141**, 182. doi:10.1088/0004-6256/141/6/182.

Mishra, S., Vivek, M., Chand, H., and Joshi, R.: 2021, *Monthly Notices of the Royal Astronomical Society* **504**, 3187. doi:10.1093/mnras/stab782.

Nair, A. and Vivek, M.: 2022, *Monthly Notices of the Royal Astronomical Society* **511**, 4946. doi:10.1093/mnras/stac204.

Shen, Y. and Ho, L.C.: 2014, *Nature* **513**, 210. doi:10.1038/nature13712.

Yong, S.Y., Webster, R.L., King, A.L., Bate, N.F., Labrie, K., and O'Dowd, M.J.: 2020, *Monthly Notices of the Royal Astronomical Society* **491**, 1320. doi:10.1093/mnras/stz3074.

Lundgren, B.F., Wilhite, B.C., Brunner, R.J., Hall, P.B., Schneider, D.P., York, D.G., and, ...: 2007, *The Astrophysical Journal* **656**, 73. doi:10.1086/510202.

Capellupo, D.M., Hamann, F., Shields, J.C., Halpern, J.P., and Barlow, T.A.: 2013, *Monthly Notices of the Royal Astronomical Society* **429**, 1872. doi:10.1093/mnras/sts427.

Vivek, M., Srianand, R., Petitjean, P., Mohan, V., Mahabal, A., and Samui, S.: 2014, *Monthly Notices of the Royal Astronomical Society* **440**, 799. doi:10.1093/mnras/stu288.

Rogerson, J.A., Hall, P.B., Ahmed, N.S., Rodríguez Hidalgo, P., Brandt, W.N., and Filiz Ak, N.: 2018, *The Astrophysical Journal* **862**, 22. doi:10.3847/1538-4357/aabfe5.

Grier, C.J., Brandt, W.N., Hall, P.B., Trump, J.R., Filiz Ak, N., Anderson, S.F., and, ...: 2016, *The Astrophysical Journal* **824**, 130. doi:10.3847/0004-637X/824/2/130.

Lu, W.-J. and Lin, Y.-R.: 2019, *The Astrophysical Journal* **887**, 178. doi:10.3847/1538-4357/ab53d7.

Black Hole Winds at All Scales
Proceedings IAU Symposium No. 378, 2024
G. Bruni, M. Diaz Trigo, S. Laha & K. Fukumura, eds.
doi:10.1017/S174392132300323X

The sub-parsec structure of AGN with VLTI / GRAVITY

Dieter Lutz[1], **Taro Shimizu**[1], **Eckhard Sturm**[1], **Daryl Santos**[1], **Ric Davies**[1], **Oliver Pfuhl**[1,2], **Jason Dexter**[1,3], **Yixian Cao**[1] **and GRAVITY Collaboration**

[1]Max-Planck-Institut für extraterrestrische Physik, Postfach 1312, 85741 Garching, Germany
email: lutz@mpe.mpg.de

[2]European Southern Observatory Karl-Schwarzschild-Straße 2, 85748 Garching, Germany

[3]Department of Astrophysics & Planetary Sciences, JILA, University of Colorado
Duane Physics Bldg., 2000 Colorado Ave, Boulder, CO 80309, USA

Abstract. The central parsec of AGN is a key region for the launching of winds, and near-infrared interferometry is a unique tool for its study. With GRAVITY at the VLT interferometer, we can now spatially resolve not just the hot dust continuum on milliarcsecond 'torus' scales through imaging but also the broad-line region (BLR) on microarcsecond scales through spectro-astrometry. We have mapped the kinematics of the BLR in seven nearby AGN, measured sizes of the hot dust for seventeen AGN, and reconstructed dust images for two AGN. BLR kinematics has allowed us to measure the BLR size and supermassive black hole mass independent of reverberation mapping. The ongoing GRAVITY+ upgrade will greatly enhance the sensitivity and sky coverage of GRAVITY, and first results demonstrate its power for AGN science at z∼2 and beyond.

Keywords. galaxies:active, galaxies:nuclei

The commissioning of the GRAVITY instrument (GRAVITY Collaboration 2017) at ESO's Very Large Telescope Interferometer has dramatically improved the sensitivity of near-infrared K-band interferometric observations. This has converted the decades old dream of interferometric studies of the broad-line region of AGN into a reality. Fig. 1 (GRAVITY Collaboration 2018) shows results for the quasar 3C273 and their fit with a broad-line region cloud model based on the model of Pancoast et al. (2014). The data are well represented by a thick almost face-on disk rotating around a supermassive black hole with $log(M_{\rm BH}/M_{\odot}) = 8.41$.

Of in total seven local universe BLRs studied in this way (GRAVITY Collaboration 2020, 2021, 2023), five are well fit with rotating thick disks. In two cases, introducing outflow motion within the disk, again following the Pancoast approach, provides an improved fit. A BLR radius-luminosity relation with directly measured radii starts to build up, with indication for high luminosity and/or high Eddington ratio AGN falling below the default $R \propto L^{0.5}$ scaling.

Combination of angular radii from GRAVITY spectroastrometry with linear radii from reverberation mapping can provide geometric distances to AGN, albeit with noticeable uncertainties (Wang et al. 2020; GRAVITY Collaboration 2021a).

Continuum emission from host dust close to the dust sublimation radius has been mapped in the prototypical Type 2 AGN NGC 1068 (GRAVITY Collaboration 2020a).

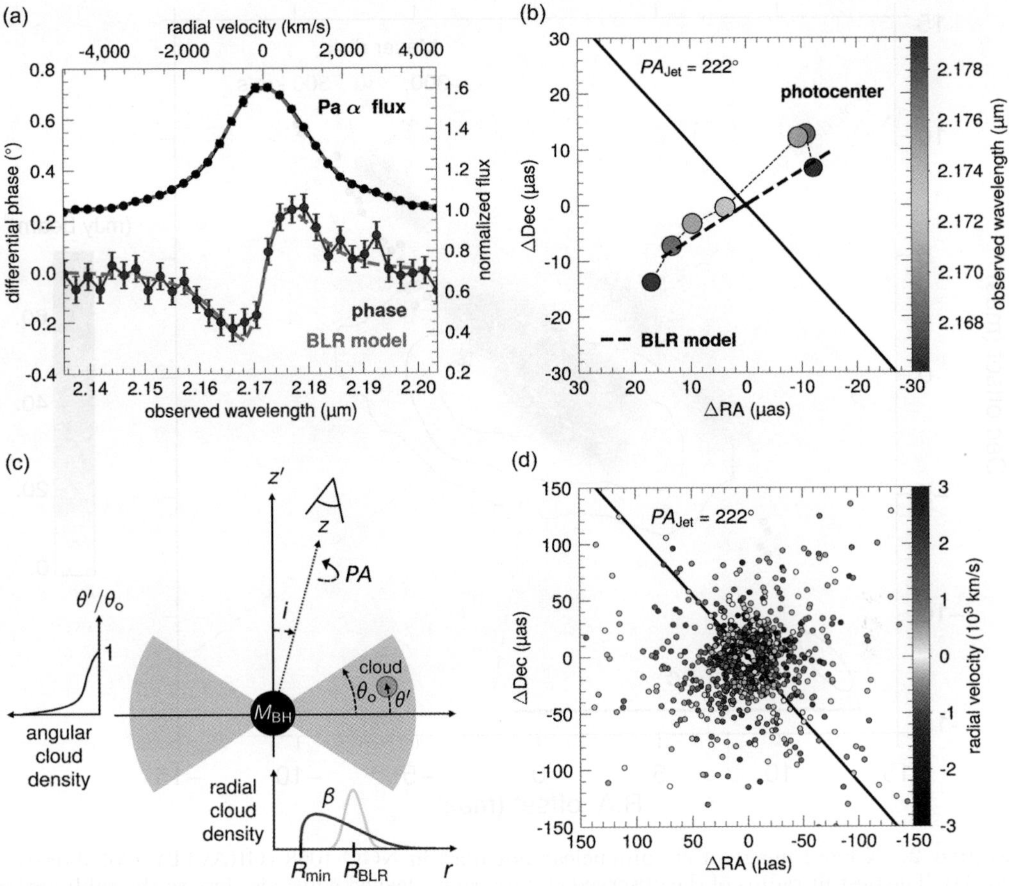

Figure 1. GRAVITY observations and modelling for 3C273 (GRAVITY Collaboration 2018). (a) Paschen α line profile and differential phases, as observed and as modelled by a thick rotating broad-line region disk. (b) Model-independent photocenters measured for different velocity slices. (c) Schematic representation of the adopted thick disk cloud model and its parameters. (d) Velocities in best fit model.

Rather than a thick torus donut, the results prefer a thin ring that is cospatial with the maser emitting clouds (Fig. 2). This clumpy ring is surrounded by thicker molecular material further out, as has been observed with ALMA. In NGC 3783, an illuminated dust cloud is visible in addition to the main K-band component (GRAVITY Collaboration 2021).

Simple hot dust sizes (rather than maps with detection over several resolution elements) have been obtained for up to now 17 local AGN (GRAVITY Collaboration 2020b, 2023a), indicating a radius-luminosity relation that is flatter than $R \propto L^{0.5}$. Such observations are quick, and a tight relation of BLR and dust size (GRAVITY Collaboration 2023b) permits the use of the inferred dust size as a proxy to BLR size in Black Hole mass measurements.

The next step in near-infrared interferometry, towards wide sky coverage, yet fainter science, and high contrast will be obtained by the GRAVITY+ project (GRAVITY+ Collaboration 2022) which will equip the VLTI 8m unit telescopes with state of the art adaptive optics, laser guide stars, and other upgrades. As a first step, the GRAVITY wide functionality is already available to the community. This makes it

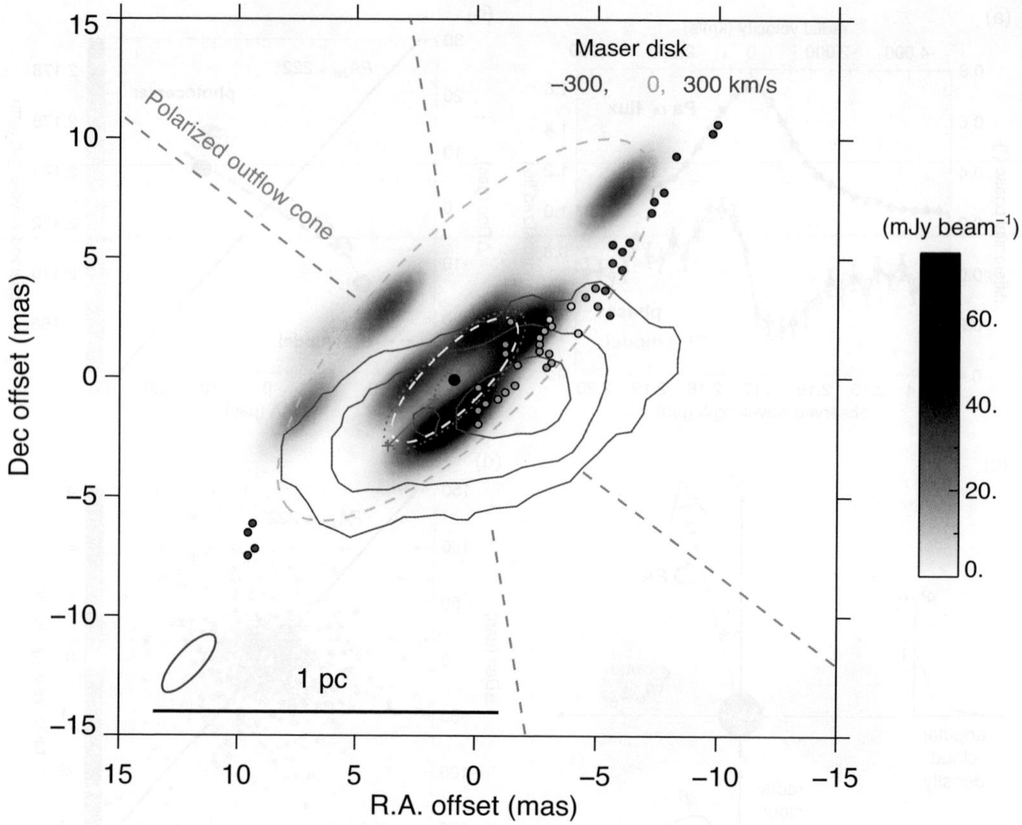

Figure 2. K-band image of circumnuclear hot dust in NGC 1068 (GRAVITY Collaboration 2020a). The best fit radius of the observed clumpy ring (dashed white) is close to the sublimation radius for hot dust (dotted orange).

possible to observe faint AGN making use of bright fringe tracking star at larger separation than before, and has already led to the first BLR detection and direct SMBH measurement of a luminous QSO at z $\sim$ 2.3 (GRAVITY+ Collaboration 2023).

References

GRAVITY Collaboration 2017, *A&A*, 602, A94
GRAVITY Collaboration 2018, *Nature*, 563, 657
GRAVITY Collaboration 2020, *A&A*, 643, A154
GRAVITY Collaboration 2020a, *A&A*, 634, A1
GRAVITY Collaboration 2020b, *A&A*, 635, A92
GRAVITY Collaboration 2021, *A&A*, 648, A117
GRAVITY Collaboration 2021a, *A&A*, 654, A85
GRAVITY Collaboration 2023, in preparation
GRAVITY Collaboration 2023a, in preparation
GRAVITY Collaboration 2023b, *A&A*, 669, A14
GRAVITY+ Collaboration 2022, *The Messenger*, 189, 17
GRAVITY+ Collaboration 2023, in preparation
Pancoast, A., Brewer, B.J., Treu, T. 2014, *MNRAS*, 445, 3305
Wang, J.-M., Songsheng, Y.-Y., Li, Y.-R., Du, P., Zhang, Z.-X. 2020, *NatAs*, 4, 517

Chapter 2:
Spectroscopic methods

Black Hole Winds at All Scales
Proceedings IAU Symposium No. 378, 2024
G. Bruni, M. Diaz Trigo, S. Laha & K. Fukumura, eds.
doi:10.1017/S1743921323003277

Testing AGN outflow and accretion models with SDSS quasar demographics

Matthew J. Temple

Instituto de Estudios Astrofísicos, Universidad Diego Portales,
Av. Ejército Libertador 441, Santiago 8370191, Chile
email: `Matthew.Temple@mail.udp.cl`

Abstract. One commonly-invoked launching mechanism for AGN outflows is radiation line driving. This mechanism depends closely on the SED of the ionizing continuum, and so is inherently linked to the structure of the accretion flow. Theories of radiation line-driven winds therefore provide testable predictions as a function of black hole (BH) mass and accretion rate. In this work we confront these predictions using the ultraviolet emission line properties of 190,000 quasars from SDSS DR17. We quantify how the shape of CIV 1549Å and the equivalent width (EW) of HeII 1640Å depend on the BH mass and Eddington ratio inferred from MgII 2800Å. The blueshift of the CIV emission line is commonly interpreted as a tracer of quasar outflows, while the HeII EW traces the strength of the 10-100 eV continuum which photo-ionizes the ultraviolet emission line regions. Above $L/L_{\rm Edd} > 0.2$, there is a strong mass dependence in both CIV blueshift and HeII EW. Large CIV blueshifts are observed only in regions with both high BH mass and high accretion rate, consistent with predictions for radiation line driven winds. The observed trends in HeII and 2 keV X-ray strength are broadly consistent with theoretical models of AGN SEDs, where the ionizing SED depends on the accretion disc temperature and the strength of the soft excess. At $L/L_{\rm Edd} < 0.2$, we find a dramatic switch in behaviour: the ultraviolet emission properties show much weaker trends, and no longer agree with SED models, hinting at changes in the structure of the broad line region. Overall the observed emission line properties are generally consistent with the radiation line driving scenario, where quasar winds are governed by the SED, which itself results from the accretion flow and hence depends on both the SMBH mass and accretion rate.

Keywords. quasars: emission lines

1. Introduction

A *quasar* is a luminous AGN, in which accretion onto the central supermassive black hole (BH) produces a large amount of electromagnetic radiation which can outshine the starlight from the host galaxy. Quasars emit across many wavebands, with a plethora of emission line features in the rest-frame ultraviolet and optical (Vanden Berk *et al.* 2001). These include both broad and narrow emission features. In the rest-frame ultraviolet, the broad lines can be further subdivided into high- and low-ionization species. The low-ionization lines are commonly observed to be symmetrical, suggestive of virialized gas surrounding the central BH. These can be used to estimate the mass of the BH, and all BH mass estimates in this presentation have been derived from the low-ionization MgII 2800Å line. In contrast, the high-ionization rest-frame ultraviolet emission lines display a diversity in their kinematic profiles (Richards *et al.* 2011; Temple *et al.* 2020, 2021a,b). Some objects show strong, symmetric line emission, while others show weaker emission which is skewed and blueshifted, suggesting ionized gas outflows along the line-of-sight to the observer. The outflow signatures seen in the CIV 1550Å line have been observed to correlate with other outflow signatures: the kinematics of [OIII] 5008Å (Coatman *et al.* 2019) and broad ultraviolet absorption features (Rankine *et al.* 2020).

1.1. *Some interesting questions*

The previous observational results outlined above naturally raise the question: *What drives the diversity of ultraviolet outflow properties in quasars?* The CIV morphology is known to correlate with luminosity, but the bolometeric luminosity is a function of both the BH mass and the normalised accretion rate $L/L_{\rm Edd}$. One can further ask: *Which of these parameters is responsible for driving the wind?*

1.2. *BH wind launching mechanisms and Quasar SEDs*

At the same time, one can also think about the launching mechanism which produces the outflows traced by the ultraviolet emission features. BH winds can be launched by thermal pressure, MHD processes, or radiation pressure on dust and/or lines. CIV outflows are likely too fast for thermal winds, and the theory of MHD processes is not yet sufficiently well developed to yield testable predictions for the behaviour of BH winds. On the other hand, radiation line driving is highly sensitive to the shape of the ionizing SED. The shape of this SED in turn depends on BH mass and accretion rate (Giustini & Proga 2019), meaning we can now test whether the demographics of CIV outflows are consistent with predictions for radiation line driven winds as a function of $M_{\rm BH}$ and $L/L_{\rm Edd}$.

Quasar SEDs can be modeled with three components (e.g. Kubota & Done 2018). These are (i) thermal emission from the multi-temperature accretion disc, (ii) hard X-ray non-thermal power-law emission from the corona, and (iii) the soft X-ray excess. Radiation line driving depends on both the flux at the line energy *and* the line opacity, which depends on the flux at the ionization edge, leading to a complicated interplay between the SED and the resulting wind.

We would like to measure the ionizing SED directly, but due to absorption from the intervening inter-galactic medium, we instead have to rely on indirect probes. HeII 1640 Å probes the 'unseen' ionizing continuum in the far ultraviolet which we can't observe directly. The relative strength of observed 2 keV emission is quantified through the $\alpha_{\rm ox}$ parameter, and probes the shape of the SED at higher energies, where the SED is sensitive to the accretion rate but plays only a minor role in photo-ionizing the ultraviolet line-emitting regions. With the data which is now available from large quasar surveys, we can compare the observed $\alpha_{\rm ox}$ and HeII strength with predictions from theoretical SED models such as those described by Kubota & Done (2018).

2. Data

The data and results presented in this talk were from a preliminary version of work which has since been published in *MNRAS* (Temple *et al.* 2023). The interested reader is encouraged to consult the *MNRAS* paper for full details. In brief, ≈190,000 quasars from the Sloan Digital Sky Survey with redshifts $1.5 < z < 2.65$ were chosen to have spectroscopic coverage of the CIV, HeII, and MgII ultraviolet emission lines. More than 5,000 of these objects also have 2 keV X-ray data available. Using a quasar SED model (Temple *et al.* 2021c) to infer 3000 Å continuum luminosities, the sample is observed to span 2.5 decades in luminosity. BH masses are estimated from the MgII line as described in Temple *et al.* (2023).

2.1. *Results*

The key results are shown in figures 3, 4, and 5 of Temple *et al.* (2023). Large CIV blueshifts only seen at $L/L_{\mathrm{Edd}} > 0.1$ and $M_{\mathrm{BH}} > 10^9 M_{\odot}$, consistent with the theories of radiation line driven winds summarized by Giustini & Proga (2019). Somewhat surprisingly, the CIV blueshift is observed to follow acute-angled wedge-shaped contours in the L/L_{Edd}–M_{BH} plane, especially given that α_{ox} does not show the same demographics. However, the observed behavior of the HeII strength matches well with the CIV blueshift, adding credence to the idea that the outflow traced by CIV is governed by the far-ultraviolet continuum.

When comparing to the Kubota & Done (2018) `qsosed` models, we see that the (admittedly somewhat naïve) zero-spin `qsosed` model matches the average α_{ox} as a function of L/L_{Edd} and M_{BH} observed in our sample. Due to an unknown covering factor, we can't compare the observed HeII strength directly with the models, but the general trend in the zero-spin `qsosed` model is consistent with HeII behaviour at $L/L_{\mathrm{Edd}} \gtrsim 0.2$. Below this threshold the HeII appears to decouple from the SED predictions, perhaps suggesting changes in the structure of the accretion flow and/or the broad-line region.

3. Conclusions

The key conclusions from this work are:

• Ultraviolet outflow signatures in quasars depend on both the BH mass *and* the accretion rate. Strong CIV outflow signatures are seen in high mass BHs with $L/L_{\mathrm{Edd}} > 0.1$, while the strongest line emission is seen in lower mass, high accretion rate objects.

• The ultraviolet emission lines depend on the strength of the ionizing SED in between directly observed X-ray and ultraviolet continuum regions. The strength of HeII emission is therefore probing complementary physics to α_{ox}.

• The emission line properties observed in the SDSS quasar population are consistent with a scenario wherein CIV traces a radiation line driven wind.

Acknowledgements

Funding received from Fondecyt Postdoctorado #3220516 and an IAU travel grant.

References

Coatman, L., Hewett, P. C., Banerji, M., Richards, G. T., Hennawi, J. F., & Prochaska, J. X. 2019, *MNRAS*, 486, 5335 (arXiv:1904.13348)

Giustini, M., & Proga, D. 2019, *A&A*, 630, A94 (arXiv:1904.07341)

Kubota, A., & Done, C. 2018, *MNRAS*, 480, 1247 (arXiv:1804.00171)

Rankine, A. L., Hewett P. C., Banerji M., & Richards, G. T. 2020, *MNRAS*, 492, 4553 (arXiv:1912.08700)

Richards, G. T., Kruczek, N. E., Gallagher, S. C., Hall, P. B., Hewett, P. C., Leighly, K. M., Deo, R. P., Kratzer, R. M., & Shen, Y. 2011, *AJ*, 141, 167 (arXiv:1011.2282)

Temple, M. J., Ferland, G. J., Rankine, A. L., Hewett, P.C., Badnell, N. R., Ballance, C. P., Del Zanna, G., & Dufresne, R. P. 2020, *MNRAS*, 496, 2565 (arXiv:2006.08617)

Temple, M. J., Banerji, M., Hewett, P. C., Rankine, A. L., & Richards, G. T. 2021a, *MNRAS*, 501, 3061 (arXiv:2012.01425)

Temple, M. J., Ferland, G. J., Rankine, A. L., Chatzikos, M., & Hewett, P. C. 2021b, *MNRAS*, 505, 3247 (arXiv:2106.01379)

Temple, M. J., Hewett, P. C., & Banerji, M. 2021c, *MNRAS*, 508, 737 (arXiv:2109.04472)

Temple, M. J., Matthews, J. H., Hewett, P. C., Rankine, A. L., Richards, G. T., Banerji, M., Ferland, G. J., Knigge, C., & Stepney, M. 2023, *MNRAS*, 523, 646 (arXiv:2301.02675)

Vanden Berk, D. E. *et al.* 2001, *AJ* 122, 549 (arXiv:astro-ph/0105231)

Black Hole Winds at All Scales
Proceedings IAU Symposium No. 378, 2024
G. Bruni, M. Diaz Trigo, S. Laha & K. Fukumura, eds.
doi:10.1017/S1743921323002922

Large-Scale Features of the CON Galaxy NGC4418 with MUSE

Clare F. Wethers[1], Susanne Aalto[1], G. C. Privon[2,3,4], F. Stanley[5], J. Gallagher[6], M. Gorski[1], S. König[1], K. Onishi[1] and C. Yang[1]

[1]Department of Space, Earth and Environment, Chalmers University of Technology, Onsala Space Observatory, 439 92 Onsala, Sweden
email: wethers@chalmers.se

[2]National Radio Astronomy Observatory, Charlottesville, VA 22903, USA

[3]Department of Astronomy, University of Florida, Gainesville, FL 32611, USA

[4]Department of Astronomy, University of Virginia, Charlottesville, VA 22904, USA

[5]IRAM, Domaine Universitaire 38406 Saint Martin d'Héres, France

[6]Wisconsin IceCube Particle Astrophysics Center, Madison, WI 53703, USA

Abstract. Compact obscured nuclei (CONs) are relatively common in the centers of local (U)LIRGs, yet their nature remains unknown. Both AGN activity and extreme nuclear starbursts have been suggested as plausible nuclear power sources. The prevalence of outflows in these systems suggest that CONs represent a key phase in the nuclear feedback cycle, in which material is ejected from the central regions of the galaxy. Here, we present results from MUSE for the confirmed local CON galaxy NGC4418. For the first time we spatially map the spectral features and kinematics of the galaxy in the optical, revealing several previously unknown structures. In particular, we discover a bilateral outflow along the minor axis, an outflowing bubble, several knot structures and a receding outflow partially obscured by the galactic disk. Based on the properties of these features, we conclude that the CON in NGC4418 is most likely powered by an AGN.

Keywords. ISM: jets and outflows, galaxies: active, galaxies: evolution

1. Introduction

Compact obscured nuclei (CONs) exist in 20-40% of nearby (ultra-) luminous infrared galaxies (ULIRGs) and are primarily characterized by their extreme nuclear column densities, $n_{H2} > 10^{25}$ cm^{-2}. This extreme attenuation renders them almost invisible at mid-IR, optical and X-ray wavelengths, which, along with their compact size ($\lesssim$100 pc), makes CONs notoriously difficult to find. To date, the only method by which these objects can be identified is via a rare transition of vibrationally excited HCN, ν_2 = 1f (HCN-VIB), which traces hot dust at high column densities (Aalto et al. 2015). Although only a handful of CONs have been confirmed, all known CONs show signatures of a past gas-rich merger and evidence for molecular outflows in HCN and/or CO. The prevalence of outflows in CONs is consistent with a heavily obscured phase in which remnant material is being ejected from the center of the galaxy. CONs are therefore critical in understanding the role of nuclear feedback in galaxy evolution, yet their nature and what powers their outflows remains unknown.

On the one hand, CONs may denote an extreme nuclear nascent starburst deeply embedded in dust. While this scenario would explain the unusual FIR and radio properties of CONs (e.g. Yun et al. 2001; Roussel et al. 2003), these starbursts are rare, and several studies have argued that dust could not obscure an extended starburst to the extent required to explain the deep absorption features observed in CONs (e.g. Roche et al. 1986; Spoon et al. 2001). This has led to the idea that CONs may instead denote a class of hidden active galactic nuclei (AGN). In such a scenario, the heavy nuclear attenuation likely arise following a rapid accretion event such as a merger or galaxy-galaxy interaction (e.g. Kocevski et al. 2015; Ricci et al. 2017; Boettcher et al. 2020). The likelihood of such interaction events have been shown to increase with (U)LIRG luminosity (Sanders & Mirabel 1996), meaning if CONs are associated with rapid gas inflow they too should preferentially exist in more luminous systems. One of the key pieces of evidence to support the AGN scenario is therefore the increased CON fraction in ULIRGs (10^{11} <L_{IR} >10^{12} L⊙; ∼40 per cent) compared to LIRGs (L_{IR} >10^{11} L⊙; ∼20 per cent).

Directly distinguishing between the starburst and hidden AGN models of CONs is challenging as many of the traditional AGN signatures in the spectra are completely erased by the extreme nuclear attenuation. We therefore rely on indirect tracers of AGN and/or starburst activity to discover what powers CONs and in turn what role they play in the coevolution of galaxies and their central super-massive black holes (SMBHs).

2. Overview

We present the first results from Wethers et al. *in prep.*, based on new targeted observations from the Multi-Unit Spectroscopic Explorer (MUSE). The work focuses on the local CON galaxy NGC4418 ($z = 0.00727$). Our key results are as follows;

1. An elongation along the minor axis: This structure is present in all of the emission features (Fig. 1), but is most noticeably traced by the Hβ and [O III] emission. The angle of the observed elongation is consistent with the minor axis outflow previously observed in NGC4418 by Ohyama et al. (2019). This outflow was found to extend to northwest of the galaxy nucleus and was catagorized as a dusty superwind due to its shock-heated emission lines and enhanced stellar NaD absorption. Unlike the outflow detected by Ohyama et al. (2019) however, we find the elongation to extend in both directions from the galaxy nucleus. Kinematic measurements (Fig. 2) reveal this structure to be strongly blueshifted in both directions with a line of sight velocity, $v_{\rm los} \gtrsim 100$ km^{-1}.

2. A shell/ bubble structure: To the north of the nucleus, we detect a bubble-like structure in the Hβ emission (Fig. 1). The structure itself appears similar to that observed in the so-called *teacup galaxy*, J1430+1339, which Gagne et al. (2014) propose to be a fading AGN candidate.

3. Knots throughout the galaxy: One of the most striking findings of this work is the presence of tens of knot structures throughout the galaxy, traced by the [O III] emission. We conclude these structures to be so-called *AGN-echos*: regions of the galaxy that have been photoionized by an AGN. This conclusion is drawn primarily from three key pieces of evidence. i) Comparisons to HST V- (F336SW) and U-band (F555W) imaging (Gallagher et al. *in prep.*) reveal no optical counterpart to the knots observed with MUSE. This counterpart would be expected if the knots were tracing regions of the galaxy ionised by young stars e.g. stellar clusters or H_{II} regions. ii) Kinematic measurements (Fig. 2) suggest at least some fraction of the knots are kinematically independent from the disk rotation of the galaxy. This rules out the possibility that the knots arise from planetary nebulae and further eliminates the stellar cluster scenario, as both would result in a similar rotational motion of the knots with the stellar disk of the galaxy. iii) Placing the knots on a Baldwin, Phillips and Terlevich (BPT) diagram (Fig. 3) reveals the knots to have higher [O III]/Hβ ratios than both nearby regions of the galaxy and H_{II} regions

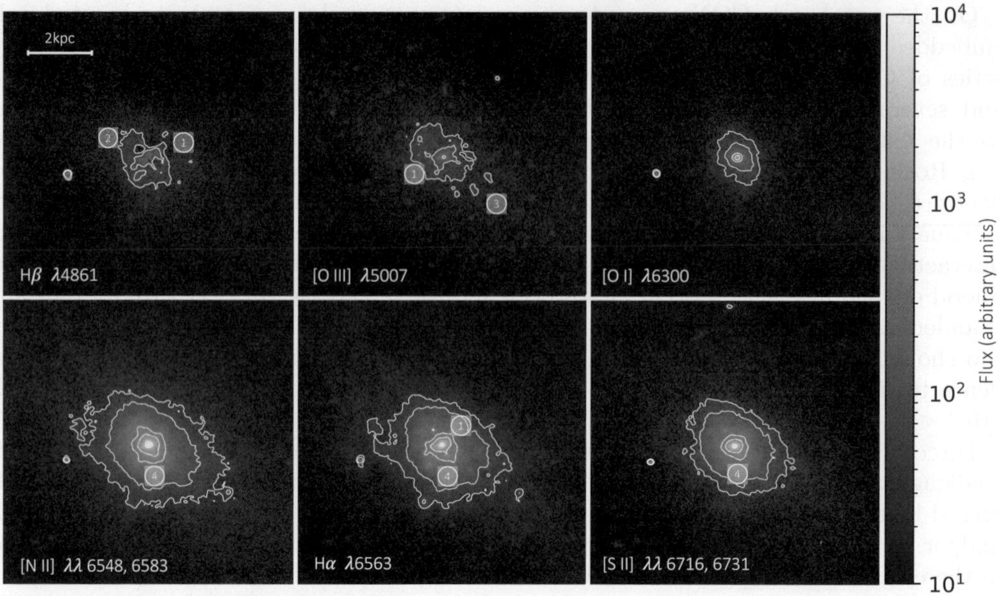

Figure 1. Integrated line maps of the key spectral features in NGC4418 with contour levels overlaid. Several prominent features of the maps are labelled (1: minor axis elongation, 2: bubble structure, 3: knots, 4: southern outflow) and are discussed in section 2.

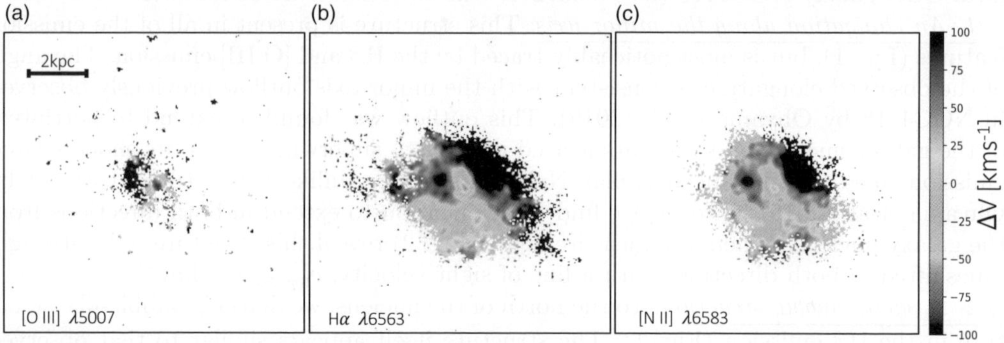

Figure 2. Moment-1 velocity maps for [O III], Hα and [N II], showing the kinematic structure of NGC4418. A 3σ signal-to-noise cut and 1σ gaussian smoothing have been applied to improve the visual output.

(Zhou et al. 2014). All the knots are found to be consistent with ionization from an AGN and/ or shock heating (Alatalo et al. 2016).

4. A southern outflow: The emission of NGC4418 shows extended emission to the south of the nucleus (Fig. 1). Kiematic measurements (Fig. 2) show this structure to be red-shifted with $v_{\rm los}$ ~50 kms^{-1}. We therefore postulate that this structure is likely an outflow orientated behind the galaxy disk.

3. Implications

The detection of multiple outflow structures in NGC4418 is consistent with the galaxy existing in an early phase of a feedback cycle in which dust is being expelled from the nuclear region via outflows. Based on the velocity and structure of the observed outflows, we suggest that it is unlikely that the CON in NGC4418 is powered solely by starburst

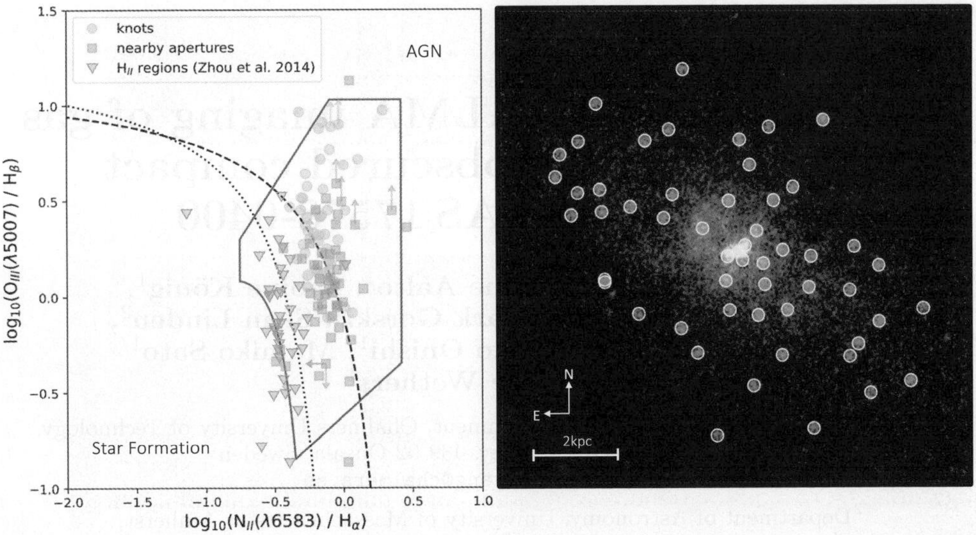

Figure 3. *Left:* BPT diagram separating ionisation from star formation and AGN. Regions are defined by the models of Kauffmann et al. (2003) (dotted) and Kewley et al. (2006) (dashed). Line ratios are measured from integrated aperture spectra (0.4 arcsec radius) centred on the knots (blue circles) and nearby galaxy regions (yellow squares), and compared to a sample of H_{II} regions (Zhou et al. 2014). The pink region shows ratios consistent with shock heating (Alatalo et al. 2016). *Right:* map of the [O III] line emission with the location of the knots overlaid.

activity. Furthermore, the knots detected in the [O III] emission are consistent with having been ionized by an AGN. We therefore conclude that the CON in NGC4418 is likely undergoing an AGN phase. Further future IFU studies are required to determine whether this classification holds for the wider CON population, or whether NGC4418 is unique in this regard. If CONs are indeed powered by AGN, this will double the number of AGN predicted to exist in the local universe (Maiolino et al. 2003) and thus have severe implications for our understanding of AGN-galaxy coevolution.

Acknowledgements: CW acknowledges support from the ERC Advanced grant 789410.

References

Aalto, S., Garcia-Burillo, S., Muller, S., et al. 2015, A&A, 574, A85
Alatalo, K., Cales, S. L., Rich, J. A., et al. 2016, ApJ Supplement Series, 224, 38
Boettcher, E., Gallagher III, J. S., Ohyama, Y., et al. 2020, A&A, 637, A17
Gagne, J. P., Crenshaw, D. M., Kraemer, S. B., et al. 2014, ApJ, 792, 72
Kauffmann, G., Heckman, T. M., Tremonti, C., et al. 2003, MNRAS, 346, 1055
Kewley, L. J., Groves, B., Kauffmann, G., & Heckman, T. 2006, MNRAS, 372, 961
Kocevski, D. D., Brightman, M., Nandra, K., et al. 2015, ApJ, 814, 104
Maiolino, R., Comastri, A., Gilli, R., et al. 2003, MNRAS, 344, L59
Ohyama, Y., Sakamoto, K., Aalto, S., & Gallagher III, J. S. 2019, ApJ, 871, 191
Ricci, C., Bauer, F. E., Treister, E., et al. 2017, MNRAS, 468, 1273
Roche, P., Aitken, D., Smith, C., & James, S. 1986, MNRAS, 218, 19P
Roussel, H., Helou, G., Beck, R., et al. 2003, ApJ, 593, 733
Sanders, D. & Mirabel, I. 1996, Annual Review of A&A, 34, 749
Spoon, H., Keane, J., Tielens, A., Lutz, D., & Moorwood, A. 2001, A&A, 365, L353
Yun, M. S., Reddy, N. A., & Condon, J. 2001, ApJ, 554, 803
Zhou, Z.-M., Wu, H., Huang, L., et al. 2014, Research in A&A, 14, 1393

Black Hole Winds at All Scales
Proceedings IAU Symposium No. 378, 2024
G. Bruni, M. Diaz Trigo, S. Laha & K. Fukumura, eds.
doi:10.1017/S1743921323003320

5–12 pc resolution ALMA imaging of gas and dust in the obscured compact nucleus of IRAS 17578-0400

Chentao Yang[1], Susanne Aalto[1], Sabine König[1], Santiago Del Palacio[1], Mark Gorski[1], Sean Linden[2], Sebastien Muller[1], Kyoko Onishi[1], Mamiko Sato[1] and Clare Wethers[1]

[1]Department of Space, Earth and Environment, Chalmers University of Technology, Onsala Space Observatory, 439 92 Onsala, Sweden
email: chentao.yang@chalmers.se

[2]Department of Astronomy, University of Massachusetts at Amherst, Amherst, MA 01003, USA

Abstract. We here present 0.02–0.04″ resolution ALMA observation of the compact obscured nucleus (CON) of IRAS 17578-0400. A dusty torus within the nucleus, approximately 4 pc in radius, has been uncovered, exhibiting a usually flat spectral index at ALMA band 3, likely due to the millimeter corona emission from the central supermassive black hole (SMBH). The dense gas disk, traced by ^{13}CO(1–0), spans 7 pc in radius and suggests an outflow driven by a disk wind due to its asymmetrical structure along the minor axis. Collimated molecular outflows (CMO), traced by the low-velocity components of the HCN(3–2) and HCO^+(3–2) lines, align with the minor axis gas disk. Examination of position-velocity plots of HCN(3–2) and HCO^+(3–2) reveals a flared dense gas disk extended a radius of ~ 60 pc, infalling and rotating at speeds of about 200 km s^{-1} and 300 km s^{-1} respectively. A centrifugal barrier, located around 4 pc from the dynamical center, implies an SMBH mass of approximately $10^8\,M_\odot$, consistent with millimeter corona emission estimates. The CMO maintains a steady rotation speed of 200 km s^{-1} over the 100 pc scale along the minor axis. The projected speed of the CMO is about 80 km s^{-1}, corresponding to around ~ 500 km s^{-1}, assuming an inclination angle of 80°. Such a kinematics structure of disk-driven collimated rotating molecular outflow with gas supplies from a falling rotating disk indicates that the feedback of the compact obscured nucleus is likely regulated by the momentum transfer of the molecular gas that connects to both the feeding of the nuclear starburst and supermassive black hole.

Keywords. galaxies: evolution, galaxies: nuclei, galaxies: active, galaxies: ISM, galaxies: kinematics and dynamics, galaxies: structure, radio continuum: galaxies

1. Introduction

Compact obscured nuclei (CONs) represent an extreme phase of galaxy evolution where rapid supermassive black hole growth and/or compact star-forming activity is completely obscured by gas and dust (e.g., Sakamoto et al. 2013; Aalto et al. 2015; González-Alfonso and Sakamoto 2019; Falstad et al. 2021). CONs are found in local luminous and ultra-luminous infrared galaxies (U/LIRGs), with an occurrence rate of 38% in ULIRGs and 21% in LIRGs (Falstad et al. 2021). These CONs are extremely dusty with high column densities that obscure X-ray to far-infrared emission and sometimes only leave optically thin windows at the submillimeter to radio bands (Scoville et al. 2017). Rotational transitions of the vibrational excited HCN thus become a good tracer of CONs, which can probe deeper into the obscured nuclei, absorbing mid-infrared photons (e.g., 14 μm, Aalto et al. 2015) and radiate at submillimeter bands. CONs are, therefore, selected based on the HCN-vib lines, considering both their luminosity and surface density. The intense infrared radiation emanating from warm dust in CONs can account for a significant fraction of the bolometric luminosity of the galaxies. This radiation is thought to be powered either by the central active galactic nuclei (AGN) and/or an extreme nuclear starburst. The exact power source of CONs is not yet fully understood (Falstad et al. 2021). Recent observations suggest that collimated molecular outflows (CMO) are ubiquitous in CONs (Barcos-Muñoz et al. 2018; Aalto et al. 2020; Gorski et al. 2023). However, the origin of the CMOs remains unclear. Understanding this could potentially shed light on the interplay between the AGN and nuclear starburst, as well as the correlation between the growth of the stellar nuclear mass and the mass of supermassive black holes (SMBHs).

To address these issues, we targeted IRAS 17578-0400 ($L_{\rm IR} = 2.3 \times 10^{11}\, L_{\odot}$, redshift $z = 0.0134$ with 284 pc/$''$), one of the LIRGs identified as a CON from the CONquest survey (Falstad et al. 2021). Using high-angular resolution ALMA Band-3 and Band-6 images, we are able to probe the spatial and kinematic structure of the ISM at scales of 5–12 pc for the first time, providing important clues about the nature of CONs.

2. Structure of the CON in IRAS 17578-0400 revealed by ALMA

We have obtained high-angular resolution ALMA observations at Band 3 ($\sim$105 GHz) and Band 6 ($\sim$255 GHz), with angular resolutions reaching 0.06-0.03$''$ and 0.04-0.02$''$(from natural weighting to uniform weighting), respectively.

Structure of the ALMA continuum.

As shown in the *upper panel* of Fig. 1, we find a very compact continuum structure from ALMA Band-3 and Band-6 images at scales of 20–30 pc. Using uniform weighting, we are able to resolve the most compact structures in the continuum in Band-6, where the source breaks into two components, with a prominent east component and a western one. These two components are also seen by the elongated structure seen in the continuum image at Band-3. Such a structure suggests either a presence of a $\sim$4 pc dusty torus around the AGN or a dual-AGN with a separation of $\sim$8 pc.

When examining the spectral energy distribution of the continuum, combined with the VJLA Ka and Ku band high-angular resolution images (Song et al. 2022), we find the SED of the CON cannot be explained by a combination of the synchrotron, free-free, and the optically thick/thin dust. The flat ALMA Band-3 in-band spectral index ($\sim$1.2) and the ratio between ALMA Band-3 and Band-6 ($\sim$3) indicate an additional contribution from the millimeter corona emission (Behar et al. 2015) is significant. Our model, derived from the fitting of the corona emission, suggests an SMBH mass of $9.1^{+2.1}_{-3.8} \times 10^7\, M_{\odot}$.

Structure of the gas components.

We highlight here the spatial structures of HCN(3–2), ^{13}CO(1–0) on top of the dust continuum in the *lower panel* of Fig. 1. As in other CONs, HCN(3–2) shows a combination of a $\sim$60 pc flared rotating disk with a $\sim$120 pc CMO (north-south direction) extended perpendicular to the disk plane on each side. It is evident that ^{13}CO shows a disk structure surrounding the dust continuum, and the velocity structure of ^{13}CO suggests that it is a rotating disk with a radius of $\sim$7 pc while the kinematic center is located in between the two dust components, supporting the scenario of a gas disk surrounding the dusty torus, rotating around the central SMBH (or a revolving binary SMBHs).

Besides the rotating disk, we also find tail-like asymmetrical outflow structures, with a blue-shifted "tail" in the northeast and a red-shifted tail towards the southwest (Fig. 1). Similar structures have been found in Galactic circumstellar disks (Alves et al. 2017).

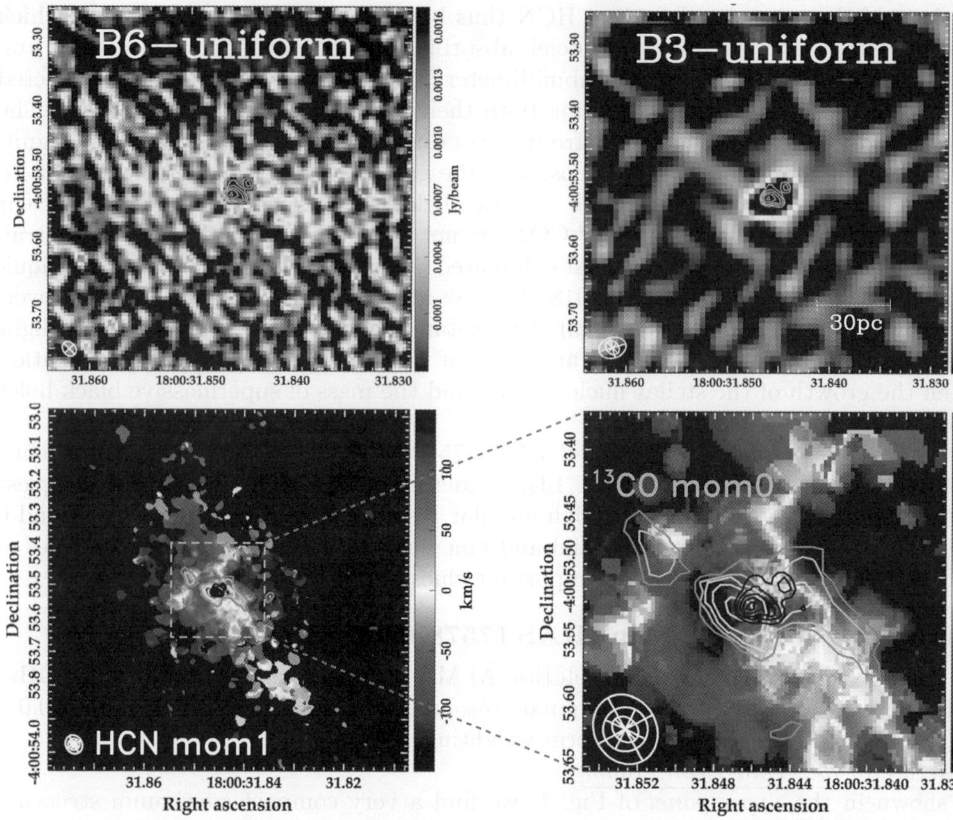

Figure 1. *Upper panels:* dust continuum in Band 6 and Band 3 from ALMA, imaged with uniform weighting to achieve the best resolution, to 0.02 and 0.03″, respectively. *Lower panels:* the moment-1 map of HCN(3–2), overlaid with yellow-green contours of the moment-0 map of ^{13}CO and the black contours of the ALMA Band-6 dust continuum.

Such an asymmetry is caused by a disk-launching wind with infalling material onto the disk asymmetrically. Considering the ^{13}CO wind is connected to the HCN(3–2) CMO, the structure of ^{13}CO indicates that the CMO is likely also driven by the rotating disk.

Kinematics.

To further understand the kinematics of the CON and CMO, we have examined the structure of the position-velocity (PV) plots of HCN(3–2), as shown in Fig. 2. The major-axis PV plot reveals bright emissions from non-Keplerian motions. We find that an infalling rotating disk model (Sakai et al. 2014) can well explain the PV structure. From the model, we derive an overall rotating speed of the flared gas disk of ~ 300 km s^{-1} with an infalling velocity of ~ 200 km s^{-1}, a centrifugal barrier radius of ~ 4 pc (where the rotation speed is ~ 300 km s^{-1}), suggesting an SMBH mass of $10^8\,M_\odot$, consistent with the estimate from millimeter corona emission. From the minor-axis PV plot, we also find that the CMO extends to about 120 pc to the minor-axis direction, which is more than half smaller than the CMO identified with a lower-spatial resolution data extending to ~ 300 pc (Falstad et al. 2021), possibly because the emission is resolved out. Offset (w.r.t the major axis) PV cuts perpendicular to the CMO direction also reveal that the CMO is rotating at the speed of ~ 200 pc with little slowing down.

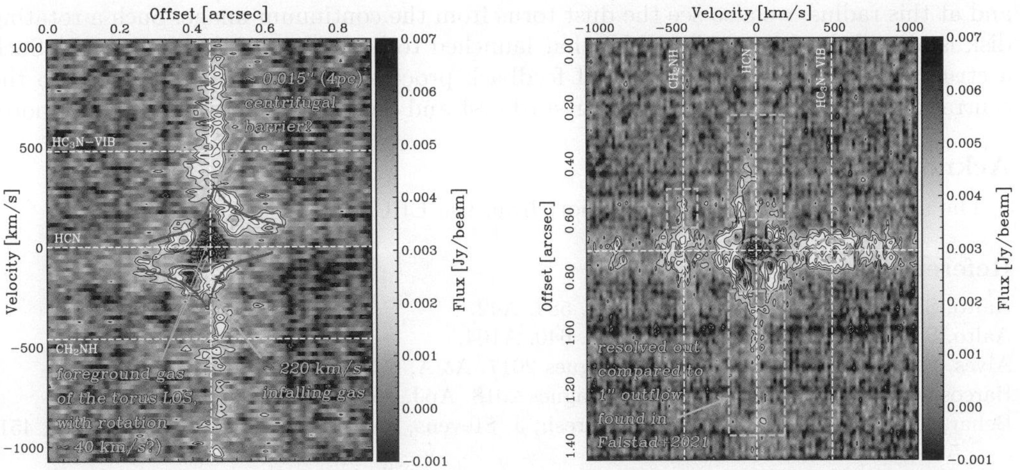

Figure 2. PV plots of HCN(3–2) of IRAS 271578-0400. The Left and right panels are extracted from the major and minor axes, respectively. We also indicate the line centers using white dashed lines. The red lines indicate a model of an infalling rotating disk.

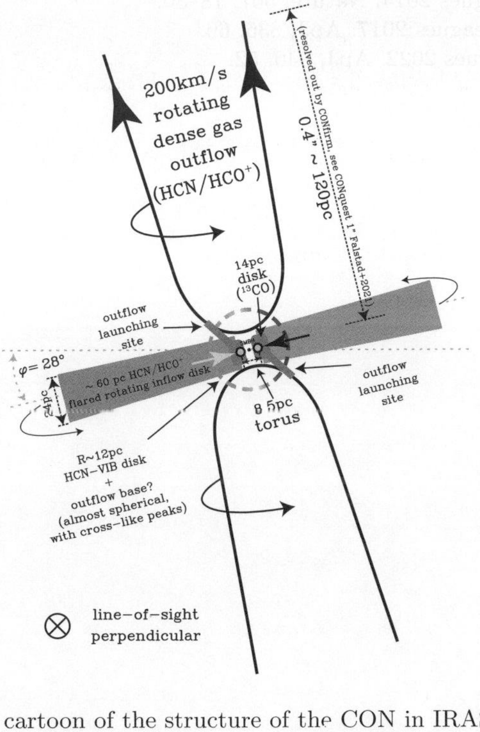

Figure 3. A cartoon of the structure of the CON in IRAS 17578-0400.

3. Conclusions and implications

Using high-spatial-resolution ALMA data, we study the spatial and kinematic structure of the compact obscured galactic nucleus of IRAS 17578-0400. We are able to resolve down to a few parsecs for the first time. Combining all the spatial and kinematic information together, we present a coherent model in Fig. 3 of the nature of the CON. We identify the key kinematic structure is dominated by a flared rotating infalling disk about 60 pc in size from the outer region towards the inner 4 pc where it reaches the centrifugal barrier, and at this radius, we also see the dust torus from the continuum image. Such a rotating disk is also likely driving the CMO that launched to both sizes of the minor axis. Such a structure might be a key driver of feedback processes, transferring the gas into the central region and feeding the nuclear starburst and SMBH, linking the growth of both.

Acknowledgments

The authors acknowledge the support from the ERC Advanced Grant 789410.

References

Aalto, S., and 19 colleagues 2015. A&A, 584, A42.
Aalto, S., and 16 colleagues 2020. A&A, 640, A104.
Alves, F. O., J. M. Girart, and 6 colleagues 2017. A&A, 603, L3.
Barcos-Muñoz, L., S. Aalto, and 7 colleagues 2018. ApJ, 853, L28.
Behar, E., R. D. Baldi, A. Laor, A. Horesh, J. Stevens, and T. Tzioumis 2015. MNRAS, 451, 517–526.
Falstad, N., and 30 colleagues 2021. A&A, 649, A105.
Gorski, M. D., S. Aalto, S. König, and 7 colleagues 2023. A&A, 670, A70.
González-Alfonso, E. and K. Sakamoto 2019. ApJ, 882, 153.
Sakamoto, K., S. Aalto, and 5 colleagues 2013. ApJ, 764, 42.
Sakai, N., and 16 colleagues 2014. Nature, 507, 78–80.
Scoville, N., and 22 colleagues 2017. ApJ, 836, 66.
Song, Y., and 25 colleagues 2022. ApJ, 940, 52.

Black Hole Winds at All Scales
Proceedings IAU Symposium No. 378, 2024
G. Bruni, M. Diaz Trigo, S. Laha & K. Fukumura, eds.
doi:10.1017/S174392132300337X

Searching for outflows in X-ray weak quasars

Bartolomeo Trefoloni[1,2]

[1]Dipartimento di Fisica e Astronomia, Università di Firenze, via G. Sansone 1, 50019 Sesto Fiorentino, Firenze, Italy

[2]INAF – Osservatorio Astrofisico di Arcetri, Largo Enrico Fermi 5, I-50125 Firenze, Italy
email: bartolomeo.trefoloni@unifi.it

Abstract. The connection between X-ray weakness and powerful X-ray outflows is both expected in a scenario where outflows are connected with radiation-driven winds, and observed in several sources, both in the local Universe and at high redshift. Here I present the first results of a new study of this possible connection based on a search for SDSS quasars with weak X-ray emission in serendipitous XMM-Newton observations. The selected objects have a "normal" optical/UV blue continuum, but a flat and extraordinarily weak X-ray spectrum. The availability of rest-frame optical/UV spectra allows to check for the signature of outflows in the absorption lines and/or in the profiles of the emission lines. This method could reveal the presence of a population of so-far overlooked outflowing quasars and confirm the connection between winds and X-ray weakness in quasars.

Keywords. quasar, AGN, outflows, X-rays, super-massive black hole

---

1. Introduction

Quasars are the most luminous persistent sources in the Universe, and there is growing evidence of their relevance for the observed properties of their host galaxies. It is general thought that an accretion disc around super-massive black holes (BH) produces their main contribution in the optical/UV (e.g., Salpeter 1964; Lynden-Bell 1969; Czerny & Elvis 1987), while the X-ray emission is likely a product of the so-called "corona" (e.g., Sunyaev & Titarchuk 1980; Haardt & Maraschi 1993), where UV disc photons are boosted via inverse Compton scattering. These two emission features are closely related (e.g. Lusso et al. 2020 and references therein), although the physics underlying such interplay is still poorly understood, and their interplay could in principle vary with the accretion parameters. At high accretion rate, for instance, the hypotheses of geometrically thin and optically thick disc (Shakura & Sunyaev 1973) could break down, with the thickening of the disc (Abramowicz et al. 1988; Chen & Wang 2004; Jiang et al. 2019).

In addition, powerful accretion-disc winds directly related to the nuclear activity (e.g., Proga 2005) could significantly alter the standard accretion process. Near Eddington sources are more likely to host of such outflows (Zubovas & King 2013; Nardini et al. 2016), which could underlie the observed relations between the black hole mass ($M_{\rm BH}$) and the galaxy properties (e.g., the $M_{\rm BH} - \sigma$ relation; Ferrarese & Merritt 2000; Gebhardt et al. 2000), although it is not yet clear whether and how AGN-driven outflows can affect their host galaxies.

At the same time, several samples of highly accreting quasars sharing quite homogeneous UV properties have recently shown an enhanced fraction of objects whose X-ray spectra are relatively flat and underluminous (by factors of $> 3 - 10$) with respect to the expectations from the $L_X - L_{UV}$ relation (e.g., Luo et al. 2015; Nardini et al. 2019; Zappacosta et al. 2020; Laurenti et al. 2022). In many cases any clear evidence for absorption has been revealed by the spectral analysis.

2. The parent sample

With the aim of expanding the $L_X - L_{UV}$ relation sample in the high luminosity and high redshift tail of the SDSS distribution, our group obtained dedicated XMM-Newton observations of 30 luminous quasars between $3.0 \leq z \leq 3.3$. This sample was selected in order to present a high degree of homogeneity in terms of UV properties, while satisfying the customary requirements of the $L_X - L_{UV}$ relation sample (e.g. Risaliti & Lusso 2019). This implied the exclusion of known broad absorption lines (BALs) objects, as well as Radio–loud ($F_{\nu,6cm}/F_{\nu,2500\mathring{A}} > 10$) and extincted ($E(B-V) > 0.1$) ones. Despite sharing such optical/UV properties, the sample presented a diverse behaviour in terms of X-ray characteristics (Nardini et al. 2019). About two thirds of the sample show X-ray luminosities in agreement with the expectations from the $L_X - L_{UV}$ relation (N quasars), and a mean photon index of $\Gamma_X \sim 1.85$, fully consistent with typical quasars at lower redshift, luminosity, and M_{BH} (e.g. Bianchi et al. 2009). Their luminosity in the 2–10 keV band spans between $4.5 \times 10^{44} \leq L_{2-10\,\mathrm{keV}} \leq 7.2 \times 10^{45}$ erg/s, being one of the most X-ray luminous samples of radio-quiet quasars ever observed. On the other hand, one third of the sources proved underluminous by factors $\sim$3–10 (W quasars). X-ray absorption at the redshift of the source is not generally statistically required by the fits of the X-ray spectra. Notwithstanding the poor quality of the data in some of cases, which leaves room for minor absorption, column densities $N_H(z) > 3 \times 10^{22}$ cm^{-2} can be ruled out.

We continued our investigation about the spectral differences between N and W quasars by performing a detailed analysis of the SDSS archival spectra for the whole sample (Lusso et al. 2021). These spectra cover the rest-frame region roughly between 900-2200 Å including the C IVλ1549 emission line. We focused on such emission feature to characterize its line properties (e.g., equivalent width, EW; line peak velocity, v_{peak}) and underlying UV continuum slope as a function of the X-ray photon index and 2–10 keV flux. We found that the composite spectrum of X-ray weak quasars is flatter ($\alpha_\lambda \sim -0.6$) than the one of X-ray normal quasars ($\alpha_\lambda \sim -1.5$). The C IV emission line is in general fainter in W quasars, but we did not report a strong blueshift (600–800 km/s) in both stacks. Such emission feature appears to be broader in the W composite spectrum (FWHM $\simeq$ 10,000 km/s) than in the N one ($\simeq$7,000 km/s), but this result is not dissimilar from other literature samples at similar redshifts (e.g. Shen et al. 2011) and luminosities (e.g. Vietri et al. 2018). We also added the sample from Timlin et al. 2020, adopting our selection criteria, with the aim to expand the dynamical range of the parameters of interest. In such a way, we confirmed the statistically significant trends of C IV v_{peak} and EW with UV luminosity at 2500 Å for both W and N, as well as the correlation between X-ray weakness and the EW of C IV. In addition, a statistically significant correlation between the hard X-ray flux and the integrated C IV flux for X-ray normal quasars was found. This relation spans more than three decades in C IV and two in terms of X-ray luminosity. X-ray weak quasars drop from the bulk of the relation by more than 0.5 dex.

Lastly, our group was awarded new dedicated observations at the Large Binocular Telescope (LBT) in the zJ and K_S bands corresponding to the rest frame Fe II$_{UV}$-Mg IIλ2798 and the Hβ-[O III] emission regions (Trefoloni et al. 2023). These new data allowed to put tighter constraints on BH masses by taking advantage of Hβ and Mg IIλ2798 single-epoch calibrations (Shen et al. 2011; Bongiorno et al. 2014), more

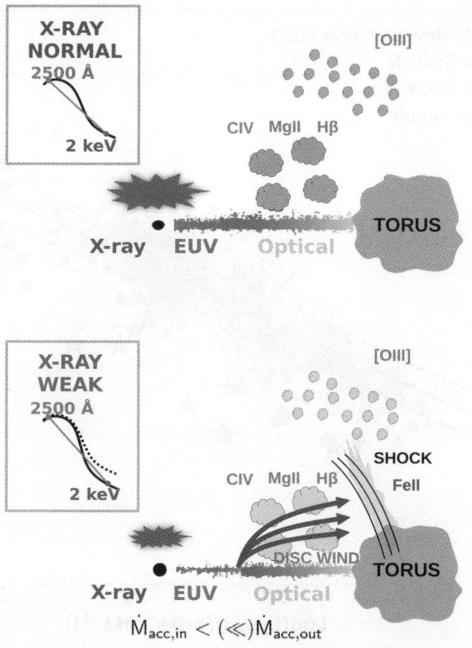

Figure 1. A sketch describing the observational features of X-ray normal (top) and weak (bottom) quasars.

reliable than the C IV ones, so far available. We found that the sample is on average made of quasars accreting near the Eddington limit $<\lambda_{Edd}>= 1.2$ ($\lambda_{Edd} = L_{bol}/L_{Edd}$, being L_{bol} and L_{Edd} respectivelythe bolometric and the Eddington luminosity), as expected for such extremely luminous objects. By fixing the slope of the continuum in order to match the one of the bluer SDSS spectral counterpart, a trend has emerged with W objects generally exhibiting more prominent Fe II_{UV} emission and, in turn, higher Fe II/Mg II ratios with respect to N ones. Such feature ratio could be enhanced in case of wind-related microturbulence and shocks in outflows (e.g. Baldwin et al. 2004; Sameshima et al. 2017; Temple et al. 2020). The EW [O III] is generally low (<20 Å) in all but one of our objects. Inclination effects are not expected to play a dominant role, since sources observed under a highly inclined line of sight with respect to the axis of the disc usually display EW [O III] $\gtrsim$ 30 Å. We also observed a lower EW of several emission lines (C IV, Hβ, [O III]) in X-ray weak quasars, and we argue that this effect could be related to the decrease of EUV photons responsible for the line production. Indeed, the $L_{[\text{O III}]}$ of the both X-ray weak and normal quasars is consistent with the high-luminosity extrapolation of the $L_{[\text{O III}]}$–$L_{2-10\,\text{keV}}$ relation from other samples in literature suggesting a common origin for the intrinsically low X-ray and [O III] emission, rather than an inclination/obscuration scenario.

We argue that, even though we do not diffusely observe unequivocal outflow footprints (e.g. large C IV and/or [O III] bluewings, BALs) in X-ray weak quasars, the presence of a mostly equatorial disc wind could explain the observational features reported so far (Figure 1). Part of the UV radiation would not be reprocessed in the X-ray corona causing intrinsic (i.e., not due to absorption) X-ray weakness. The lack of distinctive outflow signatures could be due to the modest inclination of the line of sight to the disc, consistent with the observed EW [O III] values that we find and with the huge bolometric luminosities. A higher Fe II_{UV} emission in X-ray weak quasars could be produced by

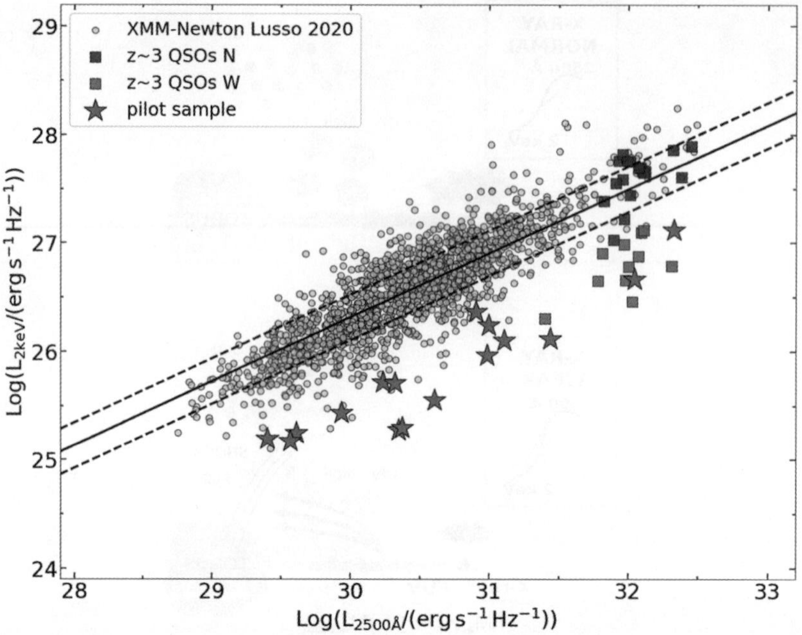

Figure 2. The $L_X - L_{UV}$ sample from Lusso et al. 2020, here used as a benchmark to select X-ray weak sources. In red are depicted the *W* sources. Large red stars represent the pilot sample, squares the z∼3 sample.

regions of microturbulence and shocks at the interface between the wind and the BLR gas. In addition, the effect of the change in the local mass accretion rate could dramatically alter the EUV SED, and ultimately determine the observed difference in terms of emission line strength (C IVλ1549, H β, [O III]). The incidence of this mechanism could be higher in samples of highly accreting sources.

3. A new pilot sample

Prompted by such results, we explored the possible correlation between the *exotic* coronal state characterized by an underluminous and flat X-ray spectra and optical/UV outflow signatures. To this purpose, we designed a new pilot sample, made of less luminous analogues of the *W* quasars, in order to investigate the tentative relation between the flat and weak X-ray emission and the possible outflowing phase. We required the sample to be X-ray underluminous by factors >3, while still presenting good X-ray and UV/optical quality (SN$\gtrsim$5) in order to be able to detect absorption and/or outflows, if present. We also required a photometric photon index $\Gamma + \delta\Gamma < 1.3$ (see Lusso et al. 2020 for more details). Since for the optical/UV side of the emission we relied on archival SDSS data, we restricted the analysis to the redshift intervals z $\leq$ 0.8 and 1.8 $\leq$ z $\leq$ 2.7 where respectively the H β-[O III]$\lambda\lambda$4959,5007 complex and the C IVλ1549-Mg IIλ2798 fall. Adopting such redshift intervals, we have also been able to retain at least one reliable BH mass virial estimator to provide an estimate of the accretion parameters (M_{BH}, λ_{Edd}). These criteria yielded a sample of 16 objects, 8 with UV spectra and 8 with optical spectra. We performed a detailed spectral analysis on X-ray and optical/UV spectra. In the X-rays we adopted a power law spectrum and included both the Galactic and the local absorption with the last one turning out to be not statistically significant in general. This meant that the observed weakness was *intrinsic* rather than due to obscuration. To perform the spectral decomposition of the optical/UV spectra, we assumed a power-law continuum and different emission line profiles (Gaussian, Lorentzian). We also allowed for a blue component in both C IVλ1549 and [O III]$\lambda\lambda$4959,5007 and included templates for the Fe II_{UV} and Fe II_{opt} pseudo-continua. Three sources out of eight with optical spectra display a prominent [O III] bluewing with offset velocity in excess of ~200 km/s and broad $\gtrsim$850 km/s profiles, likely produced in galactic-scale outflows. Half of the sources with UV spectra either show blueshifted ($\gtrsim$ 3000 km/s) or broad ($\gtrsim$2000 km/s) absorption components in the C IV profile. As a rough comparison only 2% and 3% of the sources reported in the latest SDSS catalogue (Wu & Shen 2022) exhibit blueshift in excess of 200 and 2000 km/s for the [O III] and C IV emission lines respectively†. As a side note, we should also mention that the X-ray and UV/optical spectral data have not been gathered simultaneously, thus time shift effects could also dilute or even obliterate actual correlations.

Although these preliminary results have been drawn from a relatively small pilot sample, a unifying picture relating the nuclear X-ray emission and the outer multi-scale outflow phases is emerging (e.g. Zubovas & King 2014; Cicone et al. 2018; Gaspari et al. 2020). In this framework, the weak phase of the corona could play a key role to set the environmental conditions to launch fast winds. Otherwise an efficient X-ray emission could overionize the nuclear medium, making wind line-driving unfeasible.

The results presented in these proceedings only represent a glimpse of what we believe could bridge the observed state of X-ray underluminosity to the possibly related BLR emission properties. Such multi-wavelength approach yields the potential to pave the way for a more complete description and understanding of the interplay between these two bands of the panchromatic quasar emission.

References

Abramowicz, M. A., Czerny, B., Lasota, J. P., & Szuszkiewicz, E. 1988, *ApJ*, 332, 646–658.

Baldwin, J. A., Ferland, G. J., Korista, K. T., Hamann, F., LaCluyzé, A. 2004, *ApJ*, 615(2), 610.

Bianchi, S., Guainazzi, M., Matt, G., Bonilla, N. F., Ponti, G. 2009, *A&A*, 495(2), 421–430.

Bongiorno, A., Maiolino, R., Brusa, M., Marconi, A., Piconcelli, E., Lamastra, A., Cano-Díaz, M., Schulze, A., Magnelli, B., Vignali, C., Fiore, F., Menci, N., Cresci, G., La Franca, F., Merloni, A. 2014, *MRAS*, 443(3), 2077–2091.

Chen, L. H., & Wang, J. M. 2004. *ApJ*, 614(1), 101.

Cicone, C., Brusa, M., Ramos Almeida, C., Cresci, G., Husemann, B., Mainieri, V. 2018, *Nature*, 2(3), 176–178.

Czerny, B., & Elvis, M. 1987, *ApJ*, 321, 305–320.

Ferrarese, L., & Merritt, D. 2000, *ApJ*, 539(1), L9.

Gaspari, M., Tombesi, F., Cappi, M. 2020, *Nature*, 4(1), 10–13.

Gebhardt, K., Bender, R., Bower, G., Dressler, A., Faber, S. M., Filippenko, A.V., Green, R., Grillmair, C., Ho, L.C., Kormendy, J., Lauer, T.R., Magorrian, J., Pinkney, J., Richstone, D., Tremaine, S. 2000, *ApJ*, 539(1), L13.

Haardt, F., & Maraschi, L. 1993, *ApJ*, 413, 507–517.

Jiang, Y. F., Blaes, O., Stone, J. M., Davis, S. W. 2019, *ApJ*, 885(2), 144.

Laurenti, M., Piconcelli, E., Zappacosta, L., Tombesi, F., Vignali, C., Bianchi, S., Marziani, P., Vagnetti, F., Bongiorno, A., Bischetti, M., del Olmo, A., Lanzuisi, G., Luminari, A., Middei, R., Perri, M., Ricci, C., Vietri, G. 2022, *A&A*, 657, A57.

Luo, B., Brandt, W. N., Hall, P. B., Wu, Jianfeng, Anderson, S. F., Garmire, G. P., Gibson, R. R., Plotkin, R. M., Richards, G. T., Schneider, D. P., Shemmer, O., Shen, Yue2015, The Astrophysical Journal, 805(2), 122.

† Not being the parameters of the blue component available in such catalogue, these estimates are based on the blueshift of the single broad component for the C IV and on the difference between the global profile and the core peak wavelength for the [O III].

Lusso, E., Risaliti, G., Nardini, E., Bargiacchi, G., Benetti, M., Bisogni, S., Capozziello, S., Civano, F., Eggleston, L., Elvis, M., Fabbiano, G., Gilli, R., Marconi, A., Paolillo, M., Piedipalumbo, E., Salvestrini, F., Signorini, M., Vignali, C. 2020, *A&A*, 642, A150.

Lusso, E., Nardini, E., Bisogni, S., Risaliti, G., Gilli, R., Richards, G. T., Salvestrini, F., Vignali, C., Bargiacchi, G., Civano, F., Elvis, M., Fabbiano, G., Marconi, A., Sacchi, A., Signorini, M. 2021, *A&A*, 653, A158.

Lynden-Bell, D. 1969, *Nature*, 223, 690–694.

Nardini, E., Lusso, E., Risaliti, G., Bisogni, S., Civano, F., Elvis, M., Fabbiano, G., Gilli, R., Marconi, A., Salvestrini, F., Vignali, C. 2019, *A&A*, 632, A109.

Nardini, E., Reeves, J. N., Gofford, J., Harrison, F. A., Risaliti, G., Braito, V., Costa, M. T., Matzeu, G. A., Walton, D. J., Behar, E., Boggs, S. E., Christensen, F. E., Craig, W. W., Hailey, C. J., Matt, G., Miller, J. M., O'Brien, P. T., Stern, D., Turner, T. J., Ward, M. J. 2015, *Science*, 347(6224), 860-863.

Proga, D. 2005, *ApJ*, 630(1), L9.

Risaliti, G., & Lusso, E. 2019, *Nature*, 3(3), 272–277.

Salpeter, E. E. 1964, Publications of Goddard Space Flight Center, 165.

Sameshima, H., Yoshii, Y., Kawara, K. 2017, *ApJ*, 834(2), 203.

Shakura, N. I., & Sunyaev, R. A. 1973, *A&A*, 24, 337–355.

Shen, Yue, Richards, Gordon T., Strauss, Michael A., Hall, Patrick B., Schneider, Donald P., Snedden, Stephanie, Bizyaev, Dmitry, Brewington, Howard, Malanushenko, Viktor, Malanushenko, Elena, Oravetz, Dan, Pan, Kaike, Simmons, Audrey 2011, *ApJS*, 194(2), 45.

Sunyaev, R. A., & Titarchuk, L. G. 1980, *A&A*, 86, 121–138.

Temple, Matthew J., Ferland, Gary J., Rankine, Amy L., Hewett, Paul C., Badnell, N. R., Ballance, Connor P., Del Zanna, Giulio, Dufresne, Roger P. 2020, *MNRAS*, 496(3), 2565–2576.

Timlin, J.D., Brandt, W.N., Ni, Q., Luo, B., Pu, Xingting, Schneider, D. P., Vivek, M., Yi, W. 2020, *MNRAS*, 492(1), 719–741.

Trefoloni, B., Lusso, E., Nardini, E., Risaliti, G., Bargiacchi, G., Bisogni, S., Civano, F.M., Elvis, M., Fabbiano, G.; Gilli, R., Marconi, A., Richards, G.T., Sacchi, A., Salvestrini, F., Signorini, M., Vignali, C. 2023, arXiv preprint arXiv:2305.07699.

Vietri, G., Piconcelli, E., Bischetti, M., Duras, F., Martocchia, S., Bongiorno, A., Marconi, A., Zappacosta, L., Bisogni, S., Bruni, G., Brusa, M., Comastri, A., Cresci, G., Feruglio, C., Giallongo, E., La Franca, F., Mainieri, V., Mannucci, F., Ricci, F., Sani, E., Testa, V., Tombesi, F., Vignali, C., Fiore, F. 2018, *A&A*, 617, A81.

Wang, J., Du, P., Hu, C., Netzer, H., Bai, J., Lu, K., Kaspi, S., Qiu, J., Li, Y., Wang, F.; SEAMBH Collaboration. 2014, *ApJ*, 793(2), 108.

Wu, Q., & Shen, Y. 2022, *ApJS*, 263(2), 42.

Zappacosta, L., Piconcelli, E., Giustini, M., Vietri, G., Duras, F., Miniutti, G., Bischetti, M., Bongiorno, A., Brusa, M., Chiaberge, M., Comastri, A., Feruglio, C., Luminari, A., Marconi, A., Ricci, C., Vignali, C., Fiore, F. 2020, *A&A*, 635, L5.

Zubovas, K., & King, A. 2013, *ApJ*, 769(1), 51.

Zubovas, K., & King, A. R. 2014, *MNRAS*, 439(1), 400–406.

Black Hole Winds at All Scales
Proceedings IAU Symposium No. 378, 2024
G. Bruni, M. Diaz Trigo, S. Laha & K. Fukumura, eds.
doi:10.1017/S1743921323002831

Constraining the driving mechanism of galaxy-scale winds with emission line spectra

Jonathan Stern[1], Jose Oñorbe[2], Alexander J. Richings[3], Sean D. Johnson[4] and Claude-André Faucher-Giguère[5]

[1]School of Physics & Astronomy, Tel Aviv University, Tel Aviv 69978, Israel
email: sternjon@tauex.tau.ac.il

[2]Facultad de Físicas, Universidad de Sevilla, E-41012 Seville, Spain

[3]Data Science AI and Modelling Centre, University of Hull, Hull HU6 7RX, UK

[4]Department of Astronomy, University of Michigan, Ann Arbor, MI 48109, USA

[5]Department of Physics and Astronomy and CIERA, Northwestern University, Evanston, IL 60208, USA

Abstract. Using hydrodynamic simulations and photoionization calculations, we demonstrate that quasar emission line spectra contain information on the driving mechanism of galaxy-scale outflows. Outflows driven by a hot shocked bubble are expected to exhibit LINER-like optical line ratios, while outflows driven by radiation pressure are expected to exhibit Seyfert-like line ratios. Driving by radiation pressure also has a distinct signature in the narrow UV lines, which is detected in an HST-COS spectrum of a nearby quasar hosting a large-scale wind.

Keywords. AGN, outflows, ISM: jets and outflows, Magneto-hydrdynamical simulations

---

1. Introduction

Quasar-driven galaxy-scale outflows, which are now routinely detected in emission and absorption line observations (e.g., Fiore et al. 2017), provide a potential mechanism for the central black hole to regulate star formation in the host galaxy. It is however unclear how the released black hole accretion energy couples to the host ISM and drives a large-scale outflow. Leading possibilities for this driving mechanism include direct radiation pressure from the accretion disk (e.g., Murray et al. 2005), an expanding hot bubble ($\sim 10^{10}$ K) produced when nuclear winds or jets shock against the ISM (Faucher-Giguère & Quataert 2012; Zubovas & King 2012), or pressure gradients in cosmic rays produced in the same shocks (e.g., Yusef-Zadeh et al. 2019). A method to distinguish between these possibilities would be highly beneficial for understanding the nature of these outflows and their impact on galaxy evolution (Krumholz et al. 2017; Veilleux et al. 2020).

Most existing constraints on the mass, momentum, and energy outflow rates of galaxy-scale winds are deduced from emission line spectra. *The same line spectra also contain information on the outflow driving mechanism, which has largely not been exploited.* This follows since the ionization structure of line-emitting clouds is sensitive to the confinement mechanism of the cloud's irradiated surface layer, and hence also to the wind driving mechanism.

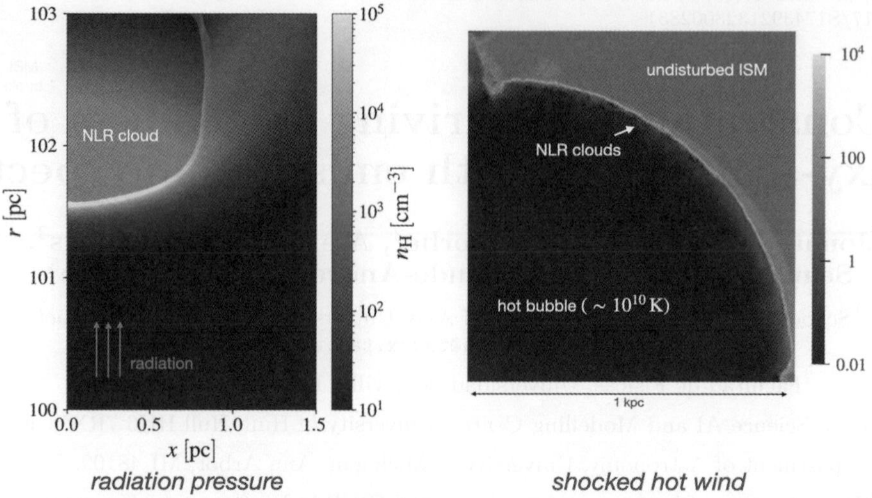

Figure 1. Simulations of NLR clouds subject to different wind-driving mechanisms. ***Left:*** an initially uniform density and spherical NLR cloud, 20 kyr after being exposed to quasar radiation (from Stern et al., in prep.). Radiation pressure both ablates the cloud and compresses the exposed surface layer, with the latter dominating the line emission. ***Right:*** Simulation of a hot bubble expanding through the host ISM (Richings & Faucher-Giguère 2018a,b). Line emission originates mainly in the compressed interface between the hot bubble and the undisturbed ISM.

In this summary we describe our efforts to infer the outflow driving mechanism from emission line spectra. We first derive quantitative predictions for line spectra assuming different driving mechanisms, and then compare these predictions with observed spectra.

2. Novel high-resolution simulations of NLR clouds subject to different outflow driving mechanisms

Figure 1 plots snapshots from the two types of hydrodynamic simulations we employ. On the left is a snapshot from a 2D axisymmetric PLUTO simulation of an ISM cloud, which is initially spherical and has a uniform density (from Stern et al., in prep.). The snapshot shows the cloud 20 kyr after being exposed to quasar radiation, which originates from the bottom of the plot. The radiation field is reduced as it proceeds within the cloud due to absorption by HI and dust grains, thus transferring both heat and momentum to the gas. The emitted radiation is derived using standard tables, and treated as an energy sink, neglecting further interactions with the gas. Spatial resolution is ≈ 0.01 pc, which is required to resolve the density structure of the cloud. The figure shows how radiation pressure both ablates the side of the cloud and compresses the surface layer against the shielded backside of the cloud. The gas pressure structure that forms in this compressed surface layer is consistent with a radiation pressure confined (RPC) slab – a 1D hydrostatic solution where the gas pressure gradient balances the pressure of the absorbed radiation (Dopita et al. 2002; Draine 2011; Stern et al. 2014; Baskin et al. 2014).

The right panel of Figure 1 plots a snapshot from the second type of simulations (from Richings & Faucher-Giguère 2018a,b), where a $\sim 10^{10}$ K bubble, as could be produced from shocks of small-scale BAL winds, expands into the host galaxy ISM. These simulations were run using the the Meshless Finite Mass hydro solver in the GIZMO code (Hopkins 2015), and have a mass resolution of ~ 30 $M_\odot$. In these simulations quasar radiation is a source of heat and ionization, radiation pressure is neglected, and radiation shielding is treated using a Sobolev-like approximation. The snapshot shows how the hot bubble compresses the ISM and drives an outflow.

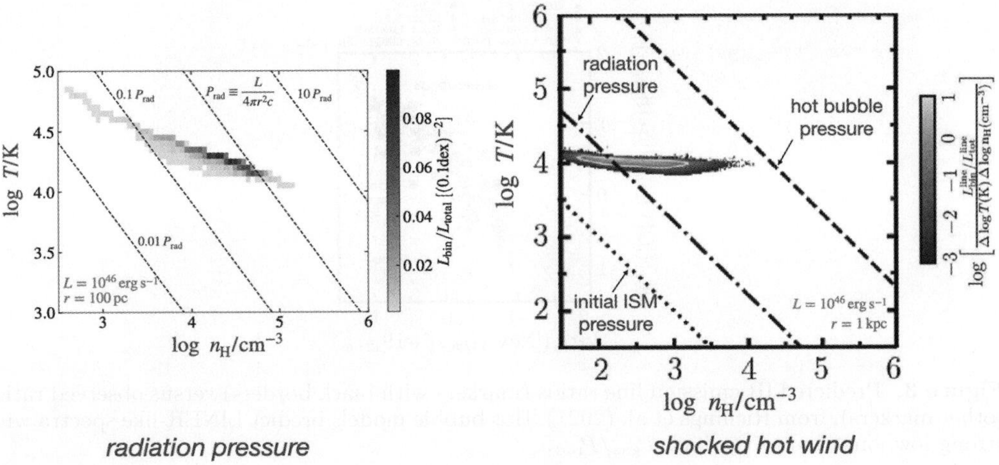

Figure 2. Density-temperature distributions weighted by Hα luminosity, in the two simulations shown in Fig. 1. Diagonal lines denote constant thermal pressures. Clouds subject to radiation pressure are compressed to $P_{\rm gas} \lesssim P_{\rm rad}$ (*left*), while in clouds subject to hot bubble pressure much of the Hα emission comes from gas with $P_{\rm gas} > P_{\rm rad}$ (*right*). Since the ratio $P_{\rm gas}/P_{\rm rad}$ sets the gas ionization level, the high $P_{\rm gas}/P_{\rm rad}$ when a hot bubble is present imply that strong low-ion lines are expected.

Figure 2 plots the density-temperature distribution of the line-emitting clouds in the two simulations, weighted by Hα luminosity. Diagonal lines denote constant thermal pressures, and specifically we mark the radiation pressure $P_{\rm rad} \equiv L/(4\pi r^2 c)$ where $L = 10^{46}$ erg s^{-1} is the bolometric luminosity and r is the cloud distance from the nucleus ($r = 100$ pc and 1 kpc in the radiation pressure and hot bubble simulations, respectively). Comparison of the two panels shows a difference in the predicted ratio of the gas pressure to the radiation pressure $P_{\rm gas}/P_{\rm rad}$. In the radiation pressure-dominated case $P_{\rm gas}/P_{\rm rad} \sim 0.3 - 1$, while in the hot bubble dominated case a higher ratio of $P_{\rm gas}/P_{\rm rad} \sim 0.5 - 10$ is evident. Compression by radiation pressure is limited to $P_{\rm rad}$, while when a dynamically important hot bubble is present clouds are compressed above this value. This result is robust to different choices of gas metallicity, spectral slope, and cloud distances.

3. Emission line predictions of different wind driving mechanisms

The relation between the driving mechanism and $P_{\rm gas}/P_{\rm rad}$ inferred above has important implications for observed emission line spectra. As discussed in Stern et al. (2016), the ratio $P_{\rm rad}/P_{\rm gas}$ sets the ionization level of the gas, and thus when a hot bubble is present and $P_{\rm gas} > P_{\rm rad}$ we expect relative strong emission from low-ionization species, in contrast with the higher-ionization lines when radiation pressure dominates and $P_{\rm rad} \lesssim P_{\rm gas}$. This difference is demonstrated in Figure 3, adapted from Richings et al. (2021). The figure shows that radiation pressure-dominated models predict relatively high ionization, Seyfert-like spectra, in contrast with lower-ionization, LINER-like spectra in hot bubble dominated models.

Emission lines from high ionization species observable in the UV provide another test of the wind driving mechanism. Radiation pressure-dominated models predict specific UV line ratios, and thus UV spectra provide a stringent test of the radiation pressure scenario. In Figure 4 (from Somalwar et al. 2020) we compare these predictions with HST-COS observations of SDSS J1356+1026 (PID: 15280, PI: Johnson), an obscured radio-quiet quasar at $z = 0.123$ which hosts a prototypical superwind. The observed line ratios are

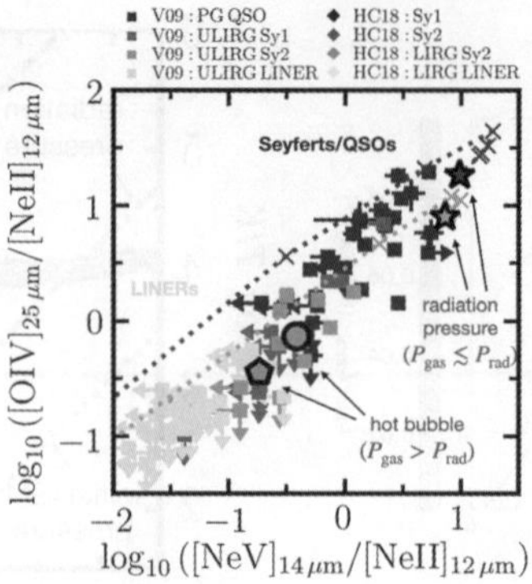

Figure 3. Predicted IR emission line ratios (markers with black borders) versus observed ratios (other markers), from Richings et al. (2021). Hot bubble models predict LINER-like spectra with strong low ions, due to the high $P_{\rm gas}/P_{\rm rad}$.

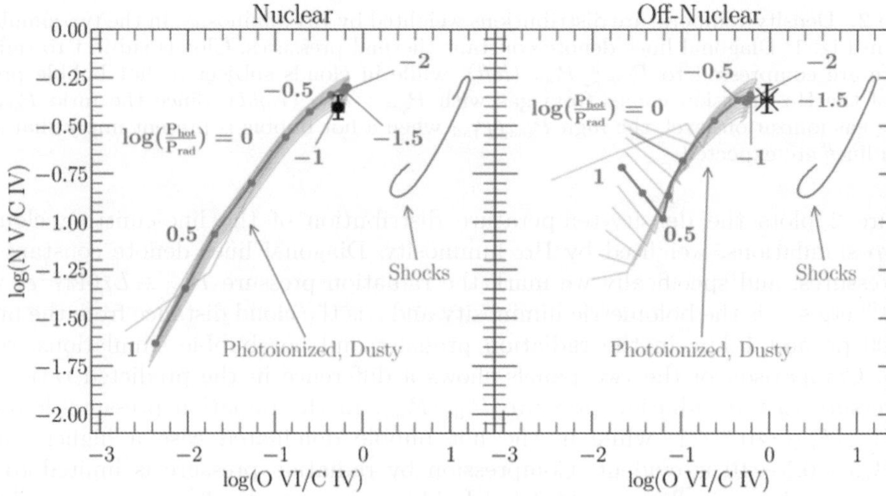

Figure 4. UV line ratios in J1356, an obscured $z = 0.1$ quasar with a prototypical superwind, compared to predictions from shock models and 1D hydrostatic photoionization models of the compressed surface layer (from Somalwar et al. 2020). The photoionization models span a range of assumed hot gas pressure to radiation pressure ratios in the range $0.01 - 10$ as marked. Different lines mark different choices for other parameters – metallicity, spectral slope and cloud distance. As $P_{\rm hot}/P_{\rm rad} \to 0$ the predicted line ratios approach asymptotic values. This limit is consistent with observed lines ratios (black markers) both in the nuclear spectrum (left) and the off-nuclear spectrum (right), *strongly suggesting radiation pressure dominates in this object.*

spot-on the predictions of the radiation pressure dominated models, strongly suggesting that radiation pressure dominates in this object. In contrast the hot bubble-dominated model predictions for high-ionization UV lines are somewhat uncertain, mainly due to the expected contribution of gas which is out of pressure, thermal and/or chemical equilibrium (Richings et al. 2021).

4. Summary and Future prospects

The above results suggest that (1) Radiation pressure driving can be identified using UV narrow line spectra; and (2) Systems with a dynamically important hot bubble are expected to have LINER-like optical and IR line ratios. It would be beneficial to search for these spectral signatures in additional objects with observed galaxy-scale winds, thus producing constraints on the wind-driving mechanism which are hard to obtain by other means. We have obtained seven additional HST-COS spectra of nearby obscured quasars for this goal (PID: 15935, PI: Johnson), and obtaining a larger sample is planned.

References

Baskin A., Laor A. & Stern J., 2014, *MNRAS*, 438, 604
Dopita M. A., Groves B. A., Sutherland R. S., Binette L. & Cecil G., 2002, *ApJ*, 572, 753
Draine B. T., 2011, *ApJ*, 732, 100
Faucher-Giguère C.-A. & Quataert E., 2012, *MNRAS*, 425, 605
Fiore F., et al., 2017, *A&A*, 601, A143
Krumholz M. R., Thompson T. A., Ostriker E. C. & Martin C. L., 2017, *MNRAS*, 471, 4061
Murray N., Quataert E. & Thompson T. A., 2005, *ApJ*, 618, 569
Richings A. J. & Faucher-Giguère C.-A., 2018a, *MNRAS*, 478, 3100
Richings A. J. & Faucher-Giguère C.-A., 2018b, *MNRAS*, 474, 3673
Richings A. J., Faucher-Giguère C.-A. & Stern J., 2021, *MNRAS*, 503, 1568
Somalwar J., Johnson S. D., Stern J., et al. 2020, *ApJL*, 890, L28
Stern J., Laor A. & Baskin A., 2014, *MNRAS*, 438, 901
Stern J., Faucher-Giguère C.-A., Zakamska N. L. & Hennawi J. F., 2016, *ApJ*, 819, 130
Veilleux S., Maiolino R., Bolatto A. D. & Aalto S., 2020, *A&AR*, 28, 2
Yusef-Zadeh F. & Wardle M., 2019, *MNRAS*, 490, L1
Zubovas K. & King A., 2012, *ApJL*, 745, L34

Chapter 3:
Feedback on the host galaxy

Black Hole Winds at All Scales
Proceedings IAU Symposium No. 378, 2024
G. Bruni, M. Diaz Trigo, S. Laha & K. Fukumura, eds.
doi:10.1017/S1743921323002818

A Multiwavelength View of Black Holes and Outflows in Post-starburst Galaxies

Yuanze Luo

William H. Miller III Department of Physics and Astronomy
Johns Hopkins University, Baltimore, MD 21218, USA
email: yluo37@jhu.edu

Abstract. Post-starburst galaxies (PSBs) have quenched (significant decline in star formation rate) both recently and rapidly ($\lesssim$1 Gyr). They are thus promising in providing insights into activities that are happening at the early stage of quenching. While studies have suggested that black hole feedback in the form of active galactic nuclei (AGN) and outflows play important roles in quenching, the details of how they impact the host galaxies and their interplay with other quenching mechanisms are still not fully understood. We find that PSBs commonly show signatures of AGN activity but they appear to be weak and/or heavily obscured. These AGN might be able to drive outflows but they are likely not strong enough to remove gas from the host galaxy. Direct evidence of AGN quenching the star formation of the host galaxy is still missing and AGN likely quench by disturbing rather than expelling the gas.

Keywords. Galaxy evolution, Active Galactic Nuclei, Post-starburst galaxy, Quenching

---

1. Introduction

Observations of a large number of galaxies have revealed their color bimodality: many galaxies are either blue and star-forming, or red and quiescent (e.g., Baldry et al. 2004). The increase in the number density of quiescent galaxies from cosmic noon to the present day (e.g., Bell et al. 2012) suggests that galaxies evolve from blue to red as star formation (SF) is quenched and gas is consumed in stars or becomes unavailable for stellar synthesis.

Of the many quenching mechanisms, the role of AGN is a heated debate. While classical cosmological simulations predict powerful AGN-driven outflows (e.g., Hopkins et al. 2008) expelling gas from galaxies to quench SF, observational evidence of galaxy quenching from AGN feedback remains circumstantial (e.g., Smethurst et al. 2017; French et al. 2018). In particular, large molecular reservoirs found in quenched galaxies (e.g., Rowlands et al. 2015) suggest that gas is stabilized against collapse into stars, possibly by disturbance from weak outflows.

Deep Balmer absorption lines in the spectra of PSBs signal the excess of intermediate-age stars formed in a recent burst that was quickly quenched. PSBs represent the fast quenching track and are ideal laboratories for studying different activities at the onset of quenching. Here we summarize our findings from several studies on the AGN and outflow activity in PSBs.

2. From Case Study to a Larger Radio-detected Sample

IC 860 is a nearby ($z\sim$0.01) PSB for which we have exquisite multiwavelength from X-ray to radio wavelengths. It has a blue optical color and is at a very early stage of transitioning to the red sequence. The in-depth study was detailed in Luo et al. (2022). In short, IC 860 is found to harbor asymmetric spiral structures that could be the remnant

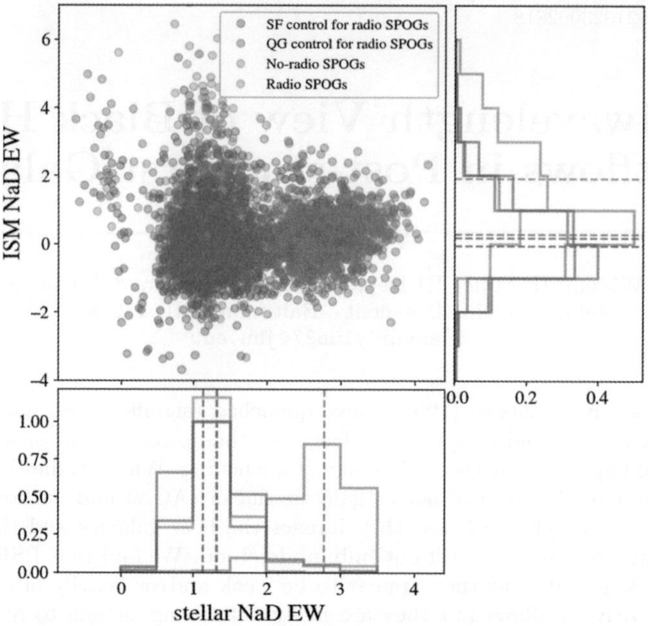

Figure 1. NaD absorption EW from ISM vs. from stellar absorption. The absorption from ISM traces cool neutral gas that could be SF fuel. Quiescent control galaxies show larger stellar NaD absorption EW as expected due to their older stellar population, while radio SPOGs show larger ISM NaD absorption EW. This fact could imply that radio SPOGs contain an elevated abundance of cool gas that is not actively forming stars, and processes associated with radio emission, e.g., AGN or small outflows, could be at play to quench SF.

of a recent merger, signatures of a weak AGN that is highly obscured, and a multiphase outflow in neutral and molecular gas that is unable to remove a significant amount of gas from the galaxy. The centrally concentrated CO molecular gas is compact (r $\sim$ 1 kpc) and kinematically misaligned with the stars. The molecular gas reservoir shows insufficient star formation efficiency, validating IC 860 as a quenching galaxy. The study implies that SF quenching and gas consumption have different timescales, and outflows prevent SF by disturbing rather than expelling the gas.

To search for AGN and outflows in a larger sample of PSBs, we turn to the Shocked POststarbust Galaxy Survey (SPOGS, Alatalo et al. 2016) which contains PSBs showing shock and/or AGN emission line ionization. Using radio emission as the smoking gun of AGN, we separate the SPOGs into radio-detected and non-radio-detected by comparing the sample to the Faint Images of the Radio Sky at Twenty Centimeters survey (FIRST, Becker et al. 1995). We also match in redshift and stellar mass to form control samples of SF and quiescent galaxies (QG). Following Luo et al. (2022), we use the NaD absorption feature at $\sim$5891 Å as a tracer of cold molecular gas and fit for its equivalent width (EW) utilizing spectra from the Sloan Digital Sky Survey (SDSS, Abazajian et al. 2009).

As shown in Fig 1, while QG show large stellar NaD EW due to old stellar populations as expected, radio-detected SPOGs show higher NaD EW from the interstellar medium (ISM) than other samples. This result suggests that there is an elevated abundance of cold gas in radio SPOGs, yet they are not actively forming stars. Processes associated with radio emission, e.g., AGN or small outflows, could be at play to quench SF. Outflows in the neutral gas, in particular, might be responsible for the larger EW we see in NaD absorption. We will look for more AGN/outflow evidence through MIR colors/spectral shape and comparison between observed radio emission and what is expected from SF.

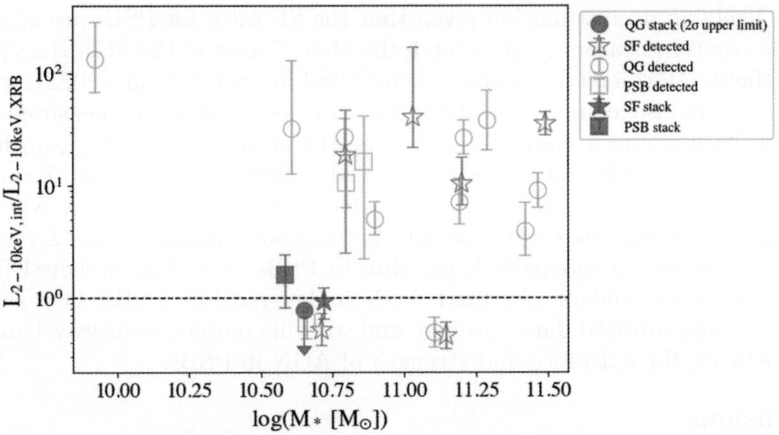

Figure 2. The ratio of the intrinsic X-ray luminosity and the expected X-ray luminosity from X-ray binaries plotted with stellar mass for both X-ray individually detected sources and the stacks of all non-detections. Individually detected PSBs show extra X-ray luminosity which could come from AGN. The stack of undetected PSBs shows a slightly enhanced ratio though still consistent with 1 considering the error bars. Given the limits of small samples, we will combine data at MIR, radio, and X-ray spectral information to constrain the AGN activity in PSBs.

This study has the potential to reveal the responsible mechanisms in galaxies at the early phase of quenching, and if outflows are confirmed, to provide a way of tracing outflows in transitioning galaxies with only radio data.

3. Exploring X-ray Observations via Stacking

While X-ray is another powerful window for probing highly energetic events like AGN, X-ray studies of PSBs have been scarce. High dust column density could greatly extinct the X-ray emission (like in IC 860), and the recent study by Lanz et al. (2022) on X-ray emission from SPOGs found that AGN, if present, have high obscuration and/or low luminosity.

X-ray sky surveys such as the eROSITA Final Equatorial Depth Survey (eFEDS, Brunner et al. 2022) allow us to perform stacking in search of faint sources that fall under the detection limit. We collected over 6000 spectral-selected PSBs from several current studies (Wild et al. 2007; Goto 2007; Pattarakijwanich et al. 2016; Alatalo et al. 2016; Tremonti et al. in prep) and 73 of them are covered by eFEDS observations. For these 73 PSBs, we match in redshift and stellar mass to control samples of SF and quiescent galaxies. We compare our samples to the detected source catalog in Brunner et al. (2022) and obtain their intrinsic luminosity estimation from X-ray spectral analysis in Liu et al. (2022). For all undetected sources, we create X-ray image cutouts and stack them following Toba et al. (2022). All luminosity measurements are converted into luminosity in the restframe 2–10 keV energy range, and we plot the ratio of the intrinsic luminosity of our galaxies to the estimated contribution from X-ray binaries (XRB, following Ito et al. 2022) in Fig 2.

X-ray-detected galaxies in different samples show some overlap in the $L_{X,int}/L_{X,XRB}$ vs. stellar mass parameter space, however, some PSBs do show extra X-ray luminosity that can not be explained by current SF. For stacked measurements, both SF galaxies and PSBs have $> 2\sigma$ detections. While both the SF and PSB stacks are consistent with $L_{X,int}/L_{X,XRB} = 1$, the PSB stack shows a slightly enhanced ratio. We note that $L_{X,XRB}$

for PSBs is likely an overestimation given that the SF rates for PSBs are often overestimated (the various indicators tend to catch the "burst" part of the SF history). Although intrinsic extinction for detected sources is corrected for in Liu et al. (2022), we still need to estimate the intrinsic column density and correct for its effect on measured luminosity in the stacks. Taking into account these factors, the plotted $L_{X,obs}/L_{X,XRB}$ here for the PSB stack is a lower limit and is likely to exceed 1 after all corrections. Expanding this stacking analysis to the full eROSITA data which covers the entire sky when available is promising in reducing the error bars and better constraining $L_{X,obs}/L_{X,XRB}$ for the stacked measurements. The excess X-ray flux in PSBs over that expected from XRBs hints at hidden weak and/or obscured AGN in low-redshift PSBs. Our samples also have radio and mid-infrared data coverage and we will combine multiwavelength data to further investigate the existence and strength of AGN in PSBs.

4. Conclusions

To trace outflows and AGN in local (z <0.2) PSBs that have recently quenched or are still quenching their SF, we expand our case study to larger samples characterized by radio detection and X-ray coverage. We found an elevated abundance of neutral gas that is not forming stars and possible neutral outflows in a radio-detected PSB sample. X-ray stacking analysis reveals possible weak AGN lurking in nearby PSBs.

Our results suggest that AGN likely play a role in the quenching of local galaxies, but whether they are the driving factor of quenching or merely collateral events as quenching happens is still not clear. While AGN are found ubiquitously in quenched galaxies at higher redshifts (e.g., Ito et al. 2022), evidence for AGN in local galaxies at the early stage of quenching remains weak or tentative. Low-redshift galaxies appear able to maintain their molecular gas reservoir with a low SF rate for several Gyr until they finally become red and gas-depleted. The power of multiwavelength data is crucial in characterizing AGN activity level in local PSBs and for testing models (e.g., Hopkins et al. 2008) predicting that the AGN phase lasts a few hundred Myr after the starburst ends to help expel remaining gas as the galaxy becomes quiescent. We have found that star formation ends not with a bang but a whimper in the local universe, and more investigation is underway to confirm this result in larger samples.

References

Abazajian, K. N., Adelman-McCarthy, J. K., Agüeros, M. A., et al. 2009, ApJS, 182, 543
Alatalo, K., Cales, S. L., Rich, J. A., et al. 2016, ApJS, 224, 38
Baldry, I. K., Glazebrook, K., Brinkmann, J., et al. 2004, ApJ, 600, 681
Becker, R. H., White, R. L., & Helfand, D. J. 1995, ApJ, 450, 559
Bell, E. F., van der Wel, A., Papovich, C., et al. 2012, ApJ, 753, 167
Brunner, H., Liu, T., Lamer, G., et al. 2022, A & A, 661, A1
French, K. D., Yang, Y., Zabludoff, A. I., & Tremonti, C. A. 2018, ApJ, 862, 2
Goto, T. 2007, MNRAS, 381, 187
Hopkins, P. F., Cox, T. J., Kereš, D., & Hernquist, L. 2008, ApJS, 175, 390
Ito, K., Tanaka, M., Miyaji, T., et al. 2022, ApJ, 929, 53
Lanz, L., Stepanoff, S., Hickox, R. C., et al. 2022, ApJ, 935, 29
Liu, T., Buchner, J., Nandra, K., et al. 2022, A & A, 661, A5
Luo, Y., Rowlands, K., Alatalo, K., et al. 2022, The Astrophysical Journal, 938, 63, aDS Bibcode: 2022ApJ...938...63L
Pattarakijwanich, P., Strauss, M. A., Ho, S., & Ross, N. P. 2016, ApJ, 833, 19
Rowlands, K., Wild, V., Nesvadba, N., et al. 2015, MNRAS, 448, 258
Smethurst, R. J., Lintott, C. J., Bamford, S. P., et al. 2017, MNRAS, 469, 3670
Toba, Y., Liu, T., Urrutia, T., et al. 2022, A & A, 661, A15
Wild, V., Kauffmann, G., Heckman, T., et al. 2007, MNRAS, 381, 543

Black Hole Winds at All Scales
Proceedings IAU Symposium No. 378, 2024
G. Bruni, M. Diaz Trigo, S. Laha & K. Fukumura, eds.
doi:10.1017/S1743921323002806

Establishing the impact of luminous AGN with multi-wavelength observations and simulations

C. M. Harrison[1], **A. Girdhar**[1,2,3] **and S. R. Ward**[2,3,4]

[1]School of Mathematics, Statistics and Physics, Newcastle University, U.K.
email: christopher.harrison@newcastle.ac.uk

[2]European Southern Observatory, Karl-Schwarzschild-Straße 2, 85748 Garching bei München, Germany

[3]Ludwig Maximilian Universität, Professor-Huber-Platz 2, 80539 Müunchen, Germany

[4]Excellence Cluster ORIGINS, Boltzmannstraße 2, 85748 Garching bei München, Germany

Abstract. Cosmological simulations fail to reproduce realistic galaxy populations without energy injection from active galactic nuclei (AGN) into the interstellar medium (ISM) and circumgalactic medium (CGM); a process called 'AGN feedback'. Consequently, observational work searches for evidence that luminous AGN impact their host galaxies. Here, we review some of this work. Multi-phase AGN outflows are common, some with potential for significant impact. Additionally, multiple feedback channels can be observed simultaneously; e.g., radio jets from 'radio quiet' quasars can inject turbulence on ISM scales, and displace CGM-scale molecular gas. However, caution must be taken comparing outflows to simulations (e.g., kinetic coupling efficiencies) to infer feedback potential, due to a lack of comparable predictions. Furthermore, some work claims limited evidence for feedback because AGN live in gas-rich, star-forming galaxies. However, simulations do not predict instantaneous, global impact on molecular gas or star formation. The impact is expected to be cumulative, over multiple episodes.

Keywords. active galactic nuclei, quasars, feedback, galaxy evolution

1. Introduction

There is a consensus across galaxy formation models and simulations that considerable energy injection, into the interstellar medium (ISM) and beyond, from active galactic nuclei (AGN) is required to regulate star formation and to reproduce many of the observable properties of massive galaxies (e.g., Stanley et al. 2015; Pillepich et al. 2018; Davé et al. 2019). This theoretical work has motivated a multitude of observational work, searching for 'direct' evidence that AGN impact upon the properties of the ISM, or circumgalactic medium (CGM), and/or the host galaxy's star formation (see e.g., Harrison 2017).

We provide a brief overview of some work studying the most luminous AGN (i.e., roughly with $L_{\rm AGN} > 10^{43}\,{\rm erg\,s^{-1}}$), which are associated with rapidly accreting supermassive black holes and that are identified mostly using optical, infrared or X-ray wavelengths. Three popular approaches that observers take to investigate if luminous AGN impact their host galaxies are discussed: (1) measuring mass outflow rates and kinetic coupling efficiencies (Section 2); (2) comparing the location of jets and outflows to spatially-resolved ISM properties and sites of star formation (Section 3); and (3) investigating the star formation rates and molecular gas properties of AGN host galaxies (Section 4). Section 5 finishes this article by discussing what can be learnt about the impact of luminous AGN from these observations, within the context of galaxy formation simulations.

2. Outflow rates and kinetic coupling efficiencies

AGN are known to drive multi-phase 'outflows' into the interstellar medium (ISM) and circumgalactic medium (CGM), as evidenced by a variety of emission-line and absorption-line studies (see e.g., Cicone et al. 2019; Harrison et al. 2018; Veilleux et al. 2020). These AGN outflows can result in increased turbulence and/or can expel material from the host galaxy's ISM (at least temporarily).

A common approach to assess the potential impact of AGN outflows is to measure their properties, including the mass outflow rates and kinetic powers. These calculations typically involve: identifying which gas is outflowing; converting fluxes to gas masses; measuring the velocities and spatial distribution of the outflowing gas; and then applying models or basic assumptions to calculate outflow rates (Harrison et al. 2018).

The inferred mass outflow rates in some sources are found to exceed the star formation rates (see discussion in e.g., Harrison 2017; Fiore et al. 2017). This means that gas is being removed more rapidly than it is forming stars, implying the potential for a significant impact on the host galaxy. However, it is very challenging to infer the long-term impact of such outflows, including understanding what fraction of this material ultimately escapes the host galaxy. Furthermore, these extreme rates are not ubiquitous. The inferred outflow rates of some more 'typical' systems can be very low to negligible (e.g., Davies et al. 2020; Ramos Almeida et al. 2022; Villar Martín et al. 2023). There are also analysis challenges, with very different results possible when applying different assumptions or using different gas tracers (e.g., Harrison et al. 2018; Davies et al. 2020).

Another common approach in the literature is to measure the ratio of outflow's kinetic power to AGN luminosity to derive the 'kinetic coupling efficiency' (see e.g., Fiore et al. 2017). In some cases, these measurements have been compared to simulations, which invoke AGN feedback to establish if the observations are consistent – or not – with feedback models. However, these comparisons are very challenging. It is important to note that the observations are typically measuring an outflow rate in one particular gas phase and often without information on how the properties vary spatially. Furthermore, the AGN luminosities are 'instantaneous' and do not contain information about earlier times, including when the outflows may have been initially launched. Therefore, these measured kinetic coupling efficiencies are not comparable quantities to the AGN coupling efficiencies invoked in sub-grid models of simulations (see discussion in Harrison et al. 2018). Indeed, if simulations track the kinetic powers of outflows as they propagate through the ISM, they can be significantly lower than their initial energy, and it remains extremely challenging to simulate the phases of outflows that are captured by observations (e.g., Costa et al. 2020).

3. Spatially-resolved investigation of impact

Recent numerical simulations have highlighted that multiple feedback 'modes' can exist simultaneously, caused by AGN-driven winds, jets and/or radiation pressure. This can include: injected turbulence to reduce star formation efficiency; removal of high-density nuclear gas (i.e., star-forming material); halting accretion from the halo (i.e., regulating future star formation); and even the localised boosting of star formation due to shocks compressing the gas to higher densities (e.g., Costa et al. 2020; Mandal et al. 2021).

Observations now also show that AGN can impact upon the ISM, and wider-scale environment, in multiple different ways. For example, luminous 'radio quiet' quasars,

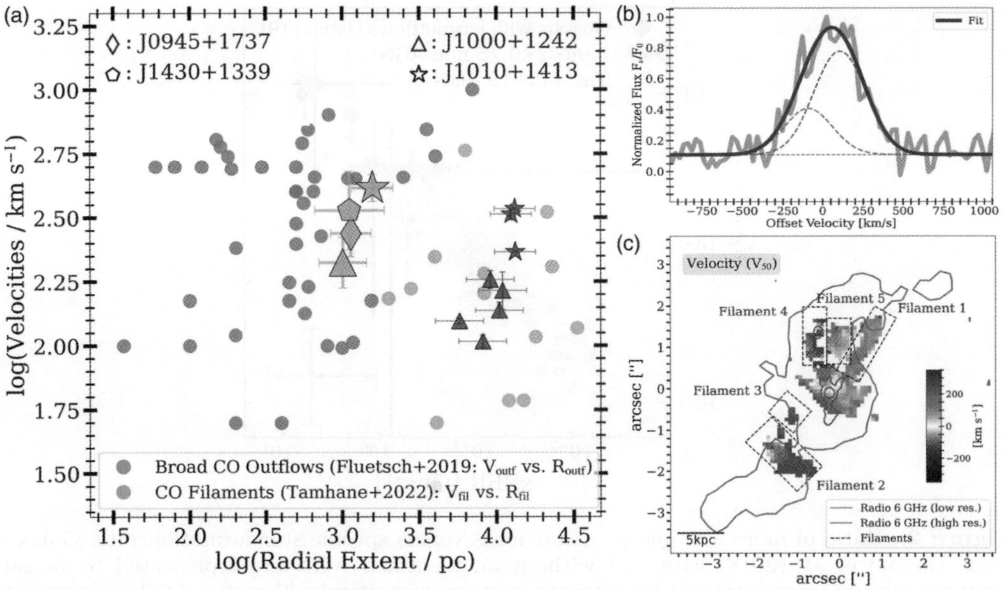

Figure 1. *(a)* Molecular gas velocities and radial extents for two types of AGN-ISM interaction, traced via CO emission: (1) broad emission-line components for a sample of AGN and star-forming galaxies (Fluetsch et al. 2019): (2) filaments around radio lobes for BCGs (Tamhane et al. 2022). The larger symbols represent data for four $z < 0.2$ 'radio quiet' quasars, two of which exhibit both types of structure. *(b)* Example CO emission-line profile for one of these quasars. *(c)* Example CO velocity map of the same quasar, with radio contours, and highlighting CO structures associated with $\approx$10 kpc radio lobes. Figures adapted from Girdhar et al. (in prep).

which have considerable radiative output and are often associated with powerful winds, are also able to interact with the star-forming material in their host galaxies via low-power radio jets and lobes (e.g., Girdhar et al. 2022; Audibert et al. 2023; Cresci et al. 2023). An example of this is showcased in Figure 1, where broad, and asymmetric CO emission-line profiles (tracing molecular outflows in the inner regions of galaxies) and molecular filaments entrained by expanding radio lobes are observed simultaneously (also see Morganti et al. 2023). These processes are typically searched for asynchronously in observations of luminous AGN (e.g., Fluetsch et al. 2019) and brightest cluster galaxies (BCGs; e.g., Russell et al. 2019; Tamhane et al. 2022), respectively.

The aforementioned observations are, at least qualitatively, similar to expectations of simulations of individual galaxies, which predict AGN will ultimately have a negative impact upon a host galaxy (e.g., Costa et al. 2020; Mandal et al. 2021). However, this global, and long-term suppression of star formation, may not easily be associated with a single AGN episode, or a localised ISM interaction. This could help explain why 'direct' observational evidence of star formation suppression continues to be scarce, tentative or inconclusive (e.g., Scholtz et al. 2021; Cresci et al. 2023). On the other hand, observations that show depleted and/or excited molecular gas within the ISM at the locations of low-power jets or outflows can indicate that the star-forming material is directly affected by the AGN (e.g., Rosario et al. 2019; Girdhar et al. 2022; Audibert et al. 2023). Nonetheless, the *long-term* and *global, galaxy-wide impact* of such processes remains more speculative. This motivates a statistical approach, as described below, that measures the global molecular gas and star formation properties of large samples of AGN host galaxies.

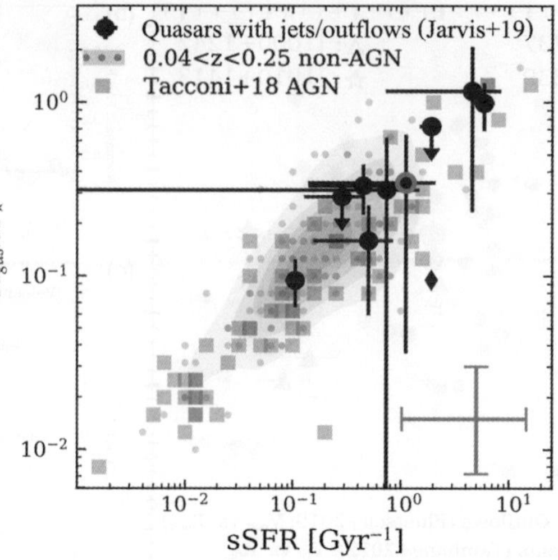

Figure 2. Ratio of molecular gas to stellar mass versus specific star formation rate. Galaxies from Tacconi et al. (2018) with and without an identified AGN are represented by density contours (also green points) and by magenta squares, respectively. Type 2 $z < 0.2$ quasars, with known outflows and/or jets are shown with black circles. Adapted from Jarvis et al. (2020).

4. Gas content and star formation rates

One popular approach to investigate the impact of AGN on their host galaxies, is to take galaxy samples and measure how the star formation properties and/or the molecular gas properties (i.e., the 'fuel' for star formation) are related (or not) to the presence of an AGN. These studies come in various forms, including: (1) investigating trends of star formation rates or molecular gas content with AGN luminosities (e.g., Stanley et al. 2015; Ramasawmy et al. 2019; Shangguan et al. 2020); and (2) establishing if AGN host galaxies have different distributions of star formation and/or molecular gas properties to non-AGN host galaxies (e.g., Scholtz et al. 2018; Koss et al. 2021; Valentino et al. 2021).

One specific example of this type of analysis is shown in Figure 2. The molecular gas fractions (derived from CO observations) are plotted as a function of specific star formation rate (i.e., star formation rate divided by stellar mass) for low redshift galaxies and AGN. The star-forming galaxies and AGN from Tacconi et al. (2018) are shown to occupy the same area of parameter space. Furthermore, quasars with known radio jets and powerful ionised outflows also have the same molecular gas and star formation properties than their less active galaxy counterparts (Jarvis et al. 2020).

With only a few exceptions (e.g., Circosta et al. 2021; Bischetti et al. 2021, which investigate molecular gas content at $z \sim 2$), studies of AGN across multiple epochs and using a variety of samples and approaches tend towards a consensus that luminous AGN typically live in gas-rich, star-forming galaxies. It can be tempting to conclude that this is in conflict with predictions that luminous AGN regulate star formation (as suggested by some observational studies), however, as we demonstrate below, this empirical result is not conflicting with predictions of cosmological simulations including AGN feedback.

5. Discussion and conclusions

We have given a very brief overview of some of the observational work investigating the impact of luminous AGN on their host galaxies (in particular the molecular gas content and the star formation). On the one hand, there is overwhelming evidence that AGN are able to modify the distribution and properties of the ISM and CGM of their host galaxies.

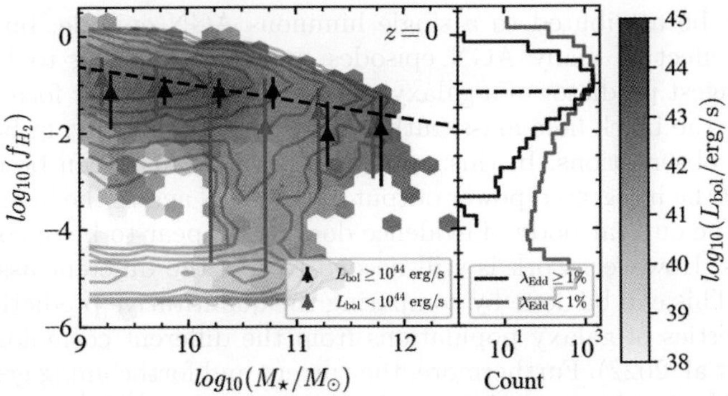

Figure 3. Ratio of molecular gas to stellar mass versus stellar mass, for $z = 0$ galaxies in the IllustrisTNG100 simulation. Colouring represents the mean AGN bolometric luminosity within each bin. Contour lines represent number density, and the triangles show the median values in bins of stellar mass for galaxies classified as AGN (blue) and non-AGN (red), based on their luminosity. The dashed line is the observed main sequence from Tacconi et al. (2018). The histograms show the distributions of gas fractions for sources classified as AGN (blue) or non-AGN (red), based on their Eddington ratios. Adapted from Ward et al. (2022).

This is due to the ever-growing body of observations that AGN radiation, winds and/or jets are able to influence the molecular gas by injecting turbulence, driving it away, and/or exciting it (Section 2 and 3). On the other hand, the *long-term* and *global significance* of these processes is less well-established, with the possibility that their impact is very localised, lasts only on short timescales, and/or is insignificant. Indeed, the observation that luminous AGN tend to live in gas-rich, star-forming galaxies has sometimes been presented as evidence against effective feedback (Section 4).

With this in mind, it is worth revisiting the cosmological simulations, which invoke AGN feedback, and motivated much of the observational work. Whilst these simulations typically lack the resolution to include a detailed physical prescription of AGN feedback, they are useful for comparing to the statistical studies of AGN and the molecular gas and star formation properties presented in Section 4. For example, Ward et al. (2022) investigated predictions from the cosmological simulations: (1) IllustrisTNG100 (Pillepich et al. 2018); (2) EAGLE (Schaye et al. 2015) and (3) SIMBA (Davé et al. 2019). These are broadly similar in scope, with $\sim$100 Mpc^3 boxes and simulation outputs that enable predictions of AGN luminosities, stellar masses, star formation rates and molecular gas content as a function of cosmic time. Ward et al. (2022) used these to reproduce the type of experiments performed by observers, as discussed in Section 4. One example of this analysis is presented in Figure 3, where the molecular gas fractions of the $z = 0$ galaxies from IllustrisTNG are presented as a function of stellar mass. The luminous AGN (selected on either luminosity or Eddington-ratio) are found to be preferentially located in gas rich galaxies. Broadly speaking, across all three simulations investigated, and at both $z = 2$ and $z = 0$, AGN are predicted to live in gas-rich, star-forming galaxies, in qualitative agreement with most observational work, although it is worth noting that there are distinct *quantitative* differences in the predictions across the three simulations due to the different feedback models used (Ward et al. 2022).

We conclude that we may not expect to have a 'smoking gun' observational signature that luminous AGN have a long-term, significant impact on molecular gas content or star formation. The reasons for this are many, with some discussion on this presented in e.g., Scholtz et al. (2018); Ward et al. (2022); Piotrowska et al. (2022). For example, an AGN may stop being visible *before* they have had an appreciable impact on their host galaxies. Furthermore, the significant impact, which ultimately results in a 'quenched' galaxy, may not be attributed to a single luminous AGN episode, but instead due to the cumulative effect of many AGN episodes. Indeed, according to Piotrowska et al. (2022), the strongest predictor of a galaxy as having quenched star formation (or not), is considered to be the black hole mass, rather than AGN luminosity, in both cosmological simulations and observations. In this context, black hole mass can be considered as an indirect tracer of the integrated power output of the AGN across the history of the galaxy.

In summary, the current body of evidence does not appear to be in conflict with AGN feedback models. However, work is still needed to test the different assumed models of AGN feedback. This can be done by comparing the quantitative predictions of the distributions of properties of galaxy populations from the different cosmological simulations (see e.g., Ward et al. 2022). Furthermore, the current and forthcoming generation of high-resolution individual galaxy simulations should make it possible to extract meaningful predictions of spatially-resolved outflow properties in different phases (e.g., mass outflow rates, kinetic powers), compare these to observations, and to assess the ultimate impact that they can have on galaxy evolution.

CMH acknowledges a United Kingdom Research and Innovation grant (MR/V022830/1). SRW acknowledges the Deutsche Forschungsgemeinschaft (DFG, German Research Foundation) under Germany's Excellence Strategy (EXC-2094-390783311).

References

Audibert A., et al., 2023, *A&A*, 671, L12
Bischetti M., et al., 2021, *A&A*, 645, A33
Cicone C., et al., 2018, *NatAs*, 2, 176
Circosta C., et al., 2021, *A&A*, 646, A96
Costa T., Pakmor R., Springel V., 2020, *MNRAS*, 497, 5229
Cresci G., et al., 2023, *A&A*, 672, A128
Davé R., et al., 2019, *MNRAS*, 486, 2827
Davies R., et al., 2020, *MNRAS*, 498, 4150
Fiore F., et al., 2017, *A&A*, 601, A143
Fluetsch A., et al., 2019, *MNRAS*, 483, 4586
Girdhar A., et al.,2022, *MNRAS*, 512, 1608
Harrison C. M., 2017, *NatAs*, 1, 0165
Harrison C. M., et al., 2018, *MNRAS*, 2, 198
Jarvis M. E., et al., 2020, *MNRAS*, 498, 1560
Koss M. J., et al., 2021, *ApJS*, 252, 29
Mandal A., et al., 2021, *MNRAS*, 508, 4738
Morganti R., Murthy S., Guillard P., Oosterloo T., Garcia-Burillo S., 2023, *Galax*, 11, 24
Pillepich A., et al., 2018, *MNRAS*, 473, 4077
Piotrowska J. M., Bluck A. F. L., Maiolino R., Peng Y., 2022, *MNRAS*, 512, 1052
Ramasawmy J., Stevens J., Martin G., Geach J. E., 2019, *MNRAS*, 486, 4320
Ramos Almeida C., et al., 2022, *A&A*, 658, A155
Rosario D. J., 2019, *ApJL*, 875, L8
Russell H. R., et al., 2019, *MNRAS*, 490, 3025
Scholtz J., et al., 2018, *MNRAS*, 475, 1288
Scholtz J., et al., 2021, *MNRAS*, 505, 5469
Shangguan J., Ho L. C., Bauer F. E., Wang R., Treister E., 2020, *ApJ*, 899, 112
Schaye J., et al., 2015, *MNRAS*, 446
Stanley F., 2015, *MNRAS*, 453, 591
Tacconi L. J., et al., 2018, *ApJ*, 853, 179
Tamhane P. D., et al., 2022, *MNRAS*, 516, 861
Valentino F., et al., 2021, *A&A*, 654, A165
Veilleux S., et al., 2020, *A&ARv*, 28, 2
Villar Martín M., et al., 2023, *A&A*, 673, A25
Ward S.R., et al.,2022, *MNRAS*, 514, 2936

Black Hole Winds at All Scales
Proceedings IAU Symposium No. 378, 2024
G. Bruni, M. Diaz Trigo, S. Laha & K. Fukumura, eds.
doi:10.1017/S1743921323003393

Ram-pressure induced radial inflows of gas to the galaxy centre

Nina Akerman[1,2], **Stephanie Tonnesen**[3], **Bianca Maria Poggianti**[1], **Rory Smith**[4] **and Antonino Marasco**[1]

[1]INAF - Astronomical Observatory of Padova, vicolo dell'Osservatorio 5, IT-35122 Padova, Italy
email: nina.akerman@inaf.it

[2]Dipartimento di Fisica e Astronomia 'Galileo Galilei', Università di Padova, vicolo dell'Osservatorio 3, IT-35122, Padova, Italy,
email: nina.akerman@studenti.unipd.it

[3]Flatiron Institute, CCA, 162 5th Avenue, New York, NY 10010, USA

[4]Departamento de Física, Universidad Técnica Federico Santa María, Vicuña Mackenna 3939, San Joaquín, Santiago de Chile

Abstract. Ram-pressure stripping (RPS) is a process known to remove gas from satellite galaxies. Recent observational studies have found an increased ratio of active galactic nuclei (AGN) among the population of RPS galaxies compared to regular galaxies in the field. To test whether ram pressure (RP) can trigger an AGN, we perform a suite of hydrodynamical wind-tunnel simulations of a massive ($M_{\rm star} = 10^{11} M_\odot$) galaxy, with inclusion of star formation, stellar feedback and high resolution up to 39 pc. We find that RP increases the inflow of gas to the galaxy centre, which in turn can result in the enhanced BH accretion, as measured by the Bondi-Hoyle model. We also estimate pressure of outflows from our accretion rates and show that AGN feedback would play an important role on the early stages of stripping, while RP itself is not so strong.

Keywords. galaxies: active, galaxies: evolution, methods: numerical

1. Introduction

In a process called ram-pressure stripping (RPS), hot intracluster medium (ICM) pushes and removes the interstellar medium (ISM) of a galaxy infalling in this cluster. Predicted by Gunn & Gott (1972), RPS has since been observed to act on galaxies in a variety of clusters (see review by Boselli et al. 2022). Apart from a more expected inevitable quenching of star formation (SF) due to gas removal, RPS has other effects on a galaxy, among which is a possible triggering of nuclear activity. Studies by Poggianti et al. (2017) and Peluso et al. (2022) show the increased number active galactic nuclei (AGN) among the population of RPS galaxies compared to regular field galaxies.

Here we will tackle the question of AGN-RPS connection through a means of simulations. Our goal is to see whether black hole (BH) accretion is enhanced enough under RP to turn a regular galaxy into an active one.

2. Methods

A detailed description of our simulation set-up is presented in Akerman et al. (2023), and here we briefly summarise it. We use an adaptive mesh refinement code *Enzo*, rendering a box region of 160 kpc on a side. An additional 5 levels of refinement allow us to achieve a maximum resolution of 39 pc.

We input a Plummer–Kuzmin stellar disk (Miyamoto & Nagai 1975) and Burkert profile for the spherical dark matter halo (Burkert 1995) as static potentials, but calculate the self-gravity of the gas and allow it to radiatively cool and form stars. For SF and stellar feedback recipes we follow Goldbaum et al. (2015, 2016). The main parameter of this SF model is the minimum threshold number density (set to $10\,\mathrm{cm}^{-3}$) at which SF can happen in a cell if its mass exceeds Jeans mass. The disk starts with metallicity $Z = 1.0 Z_{\odot}$ and the ICM has $Z = 0.3 Z_{\odot}$. In this simulation we do not include a BH seed particle, and, consequently, AGN feedback physics.

The initial conditions are based on galaxy JO201 that is both subject to RP (Bellhouse et al. 2017, 2019) and is an AGN (Poggianti et al. 2017). Several parameters go into the aforementioned static potentials, such as the stellar disk mass, set to $M_{\mathrm{star}} = 10^{11} M_{\odot}$. Parameters of the ICM wind are also based on JO201 falling into its parent cluster (Bellhouse et al. 2019), so that we model a varying wind instead of a constant one (Tonnesen 2019).

We fix a galaxy in the centre of the simulation box and add a moving ICM flowing through it, simulating a RP 'wind'. We define the wind angle as the angle between the direction of the wind and the galaxy rotation axis, and model three wind angles: 0° (face-on wind, W0), 45° (W45), and 90° (edge-on wind, W90). This allows us to study the influence of the wind angle on the centremost gas. We also simulate a fiducial galaxy not subject to RP (no wind, NW).

We let the galaxy evolve for 300 Myr before the wind reaches it. Then we separate the simulation into 4 different runs (NW, W0, W45 and W90) and model RP for 700 Myr.

3. Movement of gas in and out of the galaxy centre

With the focus of this study being BH accretion, let us first answer the question of how RP influences the movement of gas in the galaxy centre, looking closely at the innermost 500 pc. To do this, we measure mass flux though a spherical shell centred around the galaxy centre where we would expect a BH to be located, since we do not have one. The radius of this shell is 500 pc, and the mass flux is calculated as follows:

$$\dot{M} = \sum_i \frac{m_i v_i}{dL}, \tag{1}$$

where m_i is mass of the i-th cell, v_i is its radial velocity and $dL = 200$ pc is the shell's width. We measure the mass flux for cold gas, defined as having temperature $T \leq 10^{4.5}$ K, since this is the gas that might feed a BH.

We plot the results in Fig. 1, where the left and right panels present the same data, but the right one is zoomed-in to better show W0 and W45. Under RP, both inflows and outflows to the galaxy centre are increased. This manifests in the net flux as the growing difference between the peak values of the NW and RPS-galaxies. The effect is especially evident in the asymmetric W90, where the increase is about an order of magnitude during the later stages of stripping. To test for net inflow, we integrate these values and simply sum the total mass within 500 pc of the galaxy centre. We find an increased net mass inflow to, and higher total mass in, the centre of galaxies undergoing RP than the NW galaxy.

4. Black hole accretion

Now that we have shown that more cold gas inflows to the galaxy centre under RP, let us measure whether this gas can make it to the central BH. In Akerman et al. (2023) we go into detail about different BH accretion models that we can test. Here, instead,

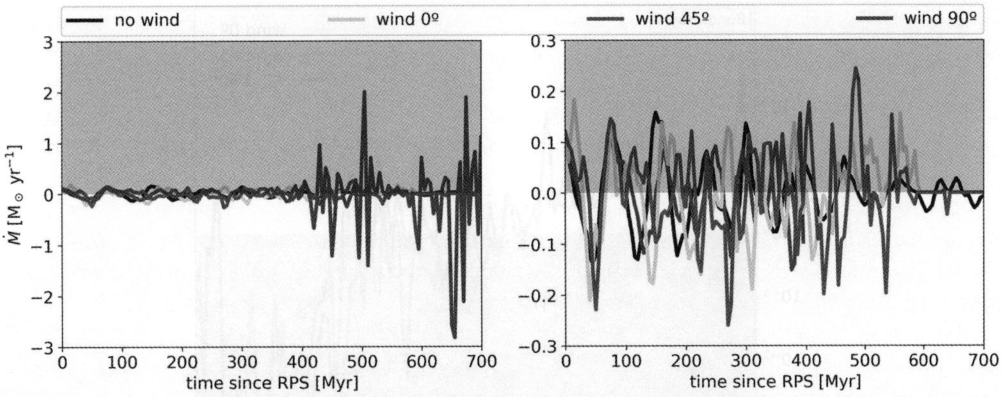

Figure 1. Net mass flux of cold gas ($T \leq 10^{4.5}$ K) though a spherical shell of radius 500 pc centred on the galaxy centre, colour-coded for different wind angles. Left and right panels present the same data, but the right one is scaled down to better show W0 and W45, and W90 is plotted only for the first 400 Myr. The grey area serves to separate inflow (negative) from outflow (positive).

we will focus on one of them, modified Bondi–Hoyle (Tremmel et al. 2017), as it gives us the highest accretion rates that will serve as an upper limit:

$$\dot{M}_{\rm BH} \propto \frac{M_{\rm BH}^2 \rho c_s}{(v_\theta^2 + c_s^2)^2}, \tag{2}$$

where $M_{\rm BH} = 10^7 M_\odot$ is assumed BH mass, ρ is the mean density and c_s is the speed of sound measured in a sphere of radius 700 pc around the galaxy centre, while v_θ is average angular velocity in a thin cylindrical shell of radii 660 pc $< R <$ 740 pc and 50 pc height. Radii of the sphere and the cylindrical shell were chosen so that the resulting accretion rate estimated by the Bondi–Hoyle model for the NW galaxy would match the accretion rate from the other two models in Akerman et al. (2023).

We find that the Bondi–Hoyle model predicts that a BH would accrete more mass under RP. RPS-galaxies also become active, defined as $\dot{M}_{\rm BH} \geq 1\%$ of the Eddington accretion limit, much more often and for longer periods of time.

In particular in face-on stripped galaxies W0 and W45, AGN feedback would 'compete' with RP in removing the gas, assuming that the former would be aligned perpendicularly to the plane of the galaxy. To estimate the relative role of AGN feedback compared to that of RP we find the energy ratio of the two forces acting on the galaxy centre. First, to measure the AGN feedback energy we follow Springel et al. (2005):

$$E_{\rm AGN} = \epsilon_f \epsilon_r \dot{M}_{\rm BH} c^2 \Delta t, \tag{3}$$

where $\epsilon_f = 0.05$ is how much of the accreted mass energy is available as feedback, $\epsilon_r = 0.1$ is radiative efficiency, c is speed of light and $\Delta t = 5$ Myr is our time step.

For the ICM wind energy, we take the kinetic energy of the wind:

$$E_{\rm ICM} = \frac{1}{2} \dot{M}_{\rm ICM} \Delta t v_{\rm ICM}^2, \tag{4}$$

where $v_{\rm ICM}$ is the wind velocity and $\dot{M}_{\rm ICM}$ is mass flux of the ICM. We measure these values in a cylinder centred 16 kpc away from the disk of W0 galaxy, with height of 2 kpc and radius of 1 kpc. We place the cylinder far enough not to catch the bow shock region but close enough so that the measured values resemble the ones immediately experienced by the galaxy disk.

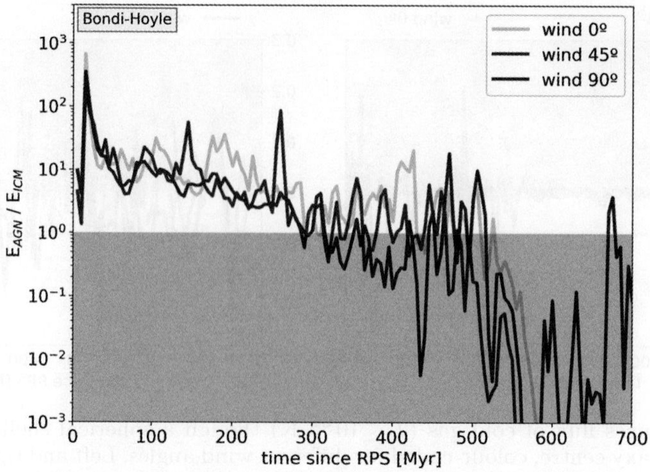

Figure 2. Ratio of energy of an AGN to kinetic energy of ICM, as a function of time, colour-coded for different wind angles. AGN energy is taken for Bondi-Hoyle accretion estimation. The grey area serves to separate AGN-dominated regime from the ICM-dominated one.

We plot the energy ratio as a function of time in Fig. 2, where $\dot{M}_{\rm BH}$ is estimated for the Bondi-Hoyle model, as it gives us the highest accretion rates. For the first 300-400 Myr the feedback energy dominates over the ICM wind, but since in our model RP increases with time, at 400-500 Myr its energy overtakes that of the AGN. This emphasises the importance of AGN feedback during the early stages of RPS. Including BH physics in the simulations could, hence, alter the galaxy evolution by stopping the ICM wind from reaching the disk at low RP.

5. Conclusions

We present the first systematic numerical study on the effect of RP on a galaxy nucleus. We find that RP increases the inflow of gas to the galaxy centre, which in turn, in accordance with observations, can result in the enhanced BH accretion. RPS-galaxies, thus, are more prone to becoming active compared to non-stripped cluster galaxies. We show that AGN feedback would play an important role on the early stages of stripping, while RP itself is not so strong. To study this effect in greater detail would require simulating RPS-galaxy with BH accretion physics and AGN feedback recipes.

Acknowledgments

This project has received funding from the European Research Council (ERC) under the European Union's Horizon 2020 research and innovation program (grant agreement No. 833824). The simulations were performed on the Frontera supercomputer operated by the Texas Advanced Computing Center (TACC) with LRAC allocation AST20007 and on the Stampede2 supercomputer operated by the Texas Advanced Computing Center (TACC) through allocation number TG-MCA06N030.

References

Akerman, N., Tonnesen, S., Poggianti B. M., et al., 2023, ApJ, 948, 18
Bellhouse, C., Jaffé, Y. L., Hau, G. K. T., et al., 2017, ApJ, 844, 49
Bellhouse, C., Jaffé, Y. L., McGee, S. L., et al., 2019, MNRAS, 485, 1157
Boselli, A., Fossati, M. & Sun, M., 2022, A&A, 30, 3
Burkert, A., 1995, ApJL, 447, L25
Goldbaum, N. J., Krumholz, M. R., Forbes, J. C., 2015, ApJ, 814, 2

Goldbaum, N. J., Krumholz, M. R., Forbes, J. C., 2015, ApJ, 827, 1
Gunn, J. E. & Gott, J. R. III, 1972, ApJ, 176, 1
Miyamoto, M. & Nagai, R., 1975, PASJ, 27, 533
Peluso, G., Vulcani, B., Poggianti, B. M., et al., 2022, ApJ, 927, 1
Poggianti, B. M., Jaffé, Y. L., Moretti, A., et al., 2017, Nature, 548, 304
Springel, V., Di Matteo, T. & Hernquist, L., 2005, MNRAS, 361, 3
Tonnesen, S., 2019, ApJ, 874, 161
Tremmel, M., Karcher, M., Governato, F., et al., 2017, MNRAS, 470, 1121

Black Hole Winds at All Scales
Proceedings IAU Symposium No. 378, 2024
G. Bruni, M. Diaz Trigo, S. Laha & K. Fukumura, eds.
doi:10.1017/S174392132300282X

Intermediate mass black hole feedback in dwarf galaxy simulations with a resolved ISM and accurate nuclear stellar dynamics

Christian Partmann[1], Thorsten Naab[1], Natalia Lahén[1], Eugene Churazov[1], Jessica M. Hislop[2], Antti Rantala[1], Peter H. Johansson[2] and Michaela Hirschmann[3]

[1]Max-Planck-Institut für Astrophysik,
Karl-Schwarzschild-Straße 1, D-85748 Garching b. München, Germany
email: partmann@mpa-garching.mpg.de

[2]Department of Physics, University of Helsinki,
Gustaf Hällströmin katu 2a, FI-00014 Helsinki, Finland

[3]Institute for Physics, Laboratory for galaxy evolution, Ecole Polytechnique Federale de Lausanne, Observatoire de Sauverny, Chemin Pegasi 51, 1290 Versoix, Switzerland

Abstract. Recent observations have established that dwarf galaxies can host black holes of intermediate mass (IMBH, $100\,\mathrm{M}_\odot < M_{\mathrm{IMBH}} \lesssim 10^5\,\mathrm{M}_\odot$). With modern numerical models, we can test the growth of IMBHs as well as their evolutionary impact on the host galaxy. Our novel subsolar-mass (0.8 solar mass) resolution simulations of dwarf galaxies ($M_* = 2\times10^7\,\mathrm{M}_\odot$) have a resolved three-phase interstellar medium and account for non-equilibrium heating, cooling, and chemistry processes. The stellar initial mass function is fully sampled between 0.08–150 $\mathrm{M}_\odot$ while massive stars can form HII regions and explode as resolved supernovae. The stellar dynamics around the IMBH is integrated accurately with a regularization scheme. We present a viscous accretion disk model for the IMBH with momentum, energy, and mass conserving wind feedback. We demonstrate how the IMBH can grow from accretion of the cold and warm gas phase and how the presence of the IMBH and its feedback impacts the gas phase structure.

Keywords. intermediate mass black holes, dwarf galaxy evolution, active galactic nuclei

---

1. Introduction

It is now established that also dwarf galaxies host massive black holes (BHs, e.g. Mezcua & Sánchez 2020). While a few supermassive BHs ($M_{\mathrm{BH}} \gtrsim 10^5\,\mathrm{M}_\odot$) have been detected in dwarf galaxies, the majority of BHs are expected to fall into the intermediate mass (IMBH) regime ($100\,\mathrm{M}_\odot < M_{\mathrm{IMBH}} \lesssim 10^5\,\mathrm{M}_\odot$, see e.g. Greene *et al.* 2020 for a review). Since gas densities in dwarfs ($M_* < 10^9\,\mathrm{M}_\odot$) are low, the potential wells shallow and the expected BH masses small, these BHs typically have low luminosities and are not necessarily found in the center of the galaxy (e.g. Sharma *et al.* 2020). As a consequence, the vast majority of IMBHs in dwarfs are expected to be unobserved and our understanding of BH occupation fractions and mass functions in this regime is still very incomplete. Furthermore, the scatter of the BH mass - stellar mass relation is significantly larger at the low mass end (e.g. Mezcua *et al.* 2023, Zaw *et al.* 2023).

Because the growth of BHs in low density environments is expected to be very limited, dwarf galaxies might give us a relatively unbiased census of the BH population in the

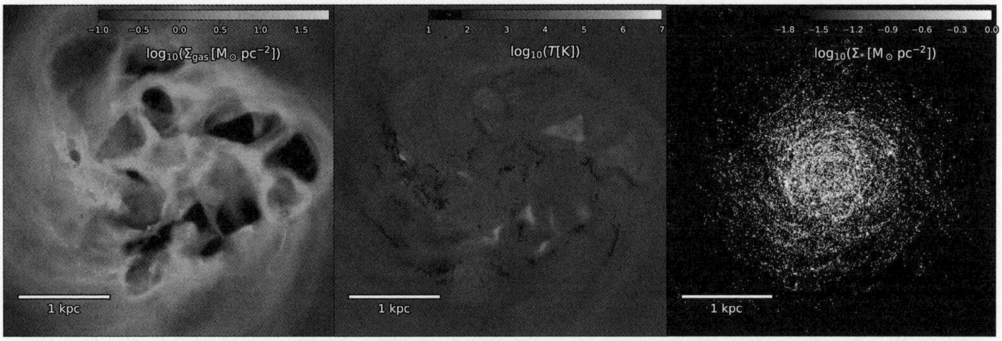

Figure 1. Face on view of the dwarf galaxy after 400 Myr of isolated evolution. The stellar feedback drives turbulence in the ISM and significantly limits the available cold gas for the BH by heating and pushing the gas out of the galactic center.

early Universe. Hence, they are well suited to explore the physics of the first BH seeds and to discriminate between proposed seeding scenarios such as direct collapse, Population III stars or gravitational runaway (Greene *et al.* 2020). In this work, we present simulations of BHs in isolated dwarf galaxies to determine typical accretion rates, cycles of activity and observational signatures of IMBHs. We focus on understanding the interplay of star formation, stellar feedback and BH physics.

2. Simulations

Model of the interstellar medium We simulate a $M_{\rm baryon} = 6 \times 10^7\,{\rm M}_\odot$ dwarf galaxy with an improved version of the smoothed-particle hydrodynamics (SPH) code Gadget3 introduced in Springel (2005) and Hu *et al.* (2014) at a gas particle resolution of $0.8\,{\rm M}_\odot$. Our model includes non-equilibrium cooling and heating for six chemical species (Hu *et al.* 2016), photo-ionizing feedback from massive stars (Hu *et al.* 2017) as well as resolved supernova explosions (Steinwandel *et al.* 2020). Star formation is based on a Jeans threshold and stars are realized as individual particles down to the Hydrogen burning limit of $0.08\,M_\odot$ as outlined in Lahén *et al.* (2023). An example for a simulated dwarf galaxy is shown in Figure 1, where the turbulent structure of the interstellar medium driven by stellar feedback is clearly visible. For this project, we have added a regularized integrator to resolve the gravitational dynamics of stars in the vicinity of the BH (Rantala *et al.* 2017). Hence, the BH does not need to be fixed to the center of the galaxy and wanders around in the central $\sim 30\,{\rm pc}$ as a result of interactions with gas clouds and close encounters with stars.

BH accretion and feedback model The accretion of gas by the BH is modeled with a sink particle prescription based on the angular momentum and Bondi radius of each SPH particle. Accreted mass is transferred to an unresolved accretion disc mass reservoir which feeds the BH on a timescale motivated by the viscous transport inside the unresolved accretion disc. This timescale is a free parameter in our model and we explore various possible parameter choices in a forthcoming paper (Partmann *et al.* in prep). We test two different BH feedback mechanisms: In the thermal feedback model, energy is injected isotropically into the interstellar medium (ISM) based on the mass accretion rate $\dot{M}_{\rm BH}$, the radiative efficiency $\epsilon_{\rm BH}$ and a coupling efficiency to the ISM $\epsilon_{\rm ISM}$

$$\dot{E}_{\rm th} = \epsilon_{\rm ISM}\epsilon_{\rm BH}\dot{M}_{\rm BH}c^2\,. \tag{2.1}$$

As an alternative, we use a kinetic feedback model similar to Ostriker *et al.* (2010), where we assume that the BH launches a collimated wind that thermalizes below the

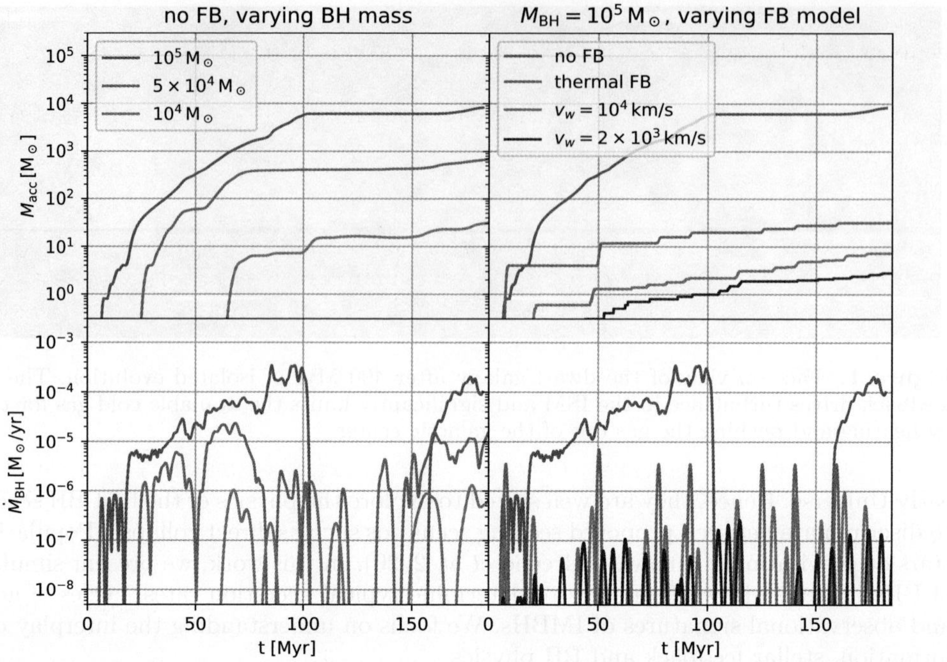

Figure 2. Accreted mass and accretion rates for varying BH masses (left) and varying feedback (FB) models (right). The BH growth is very bursty with generally low accretion rates. Stellar feedback can shut down accretion episodically even in the absence of BH feedback. The inclusion of BH feedback prohibits the growth of a BH by reducing the accretion rates by several orders of magnitude.

resolution scale of our simulation (~ 0.1 pc). Due to the energy and momentum conservation, the majority of accreted mass is re-ejected as a wind. As a result, the growth rate of the BH is smaller than the inflow of mass by a factor that depends on the assumed wind speed $v_{\rm wind}$

$$\eta = \frac{\dot{M}_{\rm outflow}}{\dot{M}_{\rm accreted}} = \frac{2\epsilon_{\rm ism}\epsilon_{\rm BH}c^2}{v_{\rm wind}^2}\,. \tag{2.2}$$

3. Results

BH growth without AGN feedback For a galaxy of stellar mass $2\times10^7\,{\rm M}_\odot$, the extrapolated scaling relations would suggest an IMBH of mass $M_{\rm BH} \lesssim 10^5\,{\rm M}_\odot$ (Greene *et al.* 2020). In the left column of Figure 2, we show the growth and accretion rates of three BHs with initial masses of $10^4, 5\times10^4$ and $10^5\,{\rm M}_\odot$ in simulations without BH feedback. The accretion rates are generally low and typically do not exceed Eddington ratios of 0.1. The growth history is bursty due to the accretion of cold clouds and the strong effect of stellar feedback, that can evacuate the dwarf galaxy center episodically or heat the gas to high temperatures where accretion is inefficient. While our heaviest tested BH can grow by $\sim 5\%$ during the simulation time, the $M_{\rm BH} = 10^4\,M_\odot$ BH shows almost no growth.

BH growth with AGN feedback Simulations without BH feedback are relevant to derive the amount of cold gas that can in principle feed the BH. However, they neglect the effect of the BH on its environment and the self-regulating nature of the AGN feedback. These effects are known to be important even in dwarf galaxies (see e.g. Schutte *et al.* 2022, Koudmani *et al.* 2022, Silk 2017). Using the $M_{\rm BH} = 10^5\,{\rm M}_\odot$ BH as a test case, we find that any type of feedback shuts down BH growth almost completely. While the thermal feedback model limits accretion by efficiently heating the ISM around the BH, the kinetic feedback models (shown for wind speeds of $v_w = 10^4$ km/s and 2×10^3 km/s) lead to even lower accretion rates, because most mass is directly ejected as a wind instead of being accreted onto the BH. We observe only a small impact on the star formation rates in the center of the dwarf galaxy.

4. Discussion

Our simulations show that the growth of an IMBH in a $M_* = 2 \times 10^7$ $M_\odot$, actively star forming ($\sim 10^{-3}$ $M_\odot$/yr) dwarf galaxy is very limited. Stellar feedback can quench the accretion for extended periods of time by heating and pushing gas out of the galactic center. In addition, the BH feedback makes BH growth even more inefficient. The resulting X-ray luminosities are very low ($\lesssim 10^{40}$ergs/s), slightly below the likely incomplete observations of local dwarfs as e.g. presented in Birchall *et al.* (2020). We conclude that more extreme environments (e.g. galaxy mergers) or less efficient feedback channels are required to grow IMBHs in dwarf galaxies. In a forthcoming paper (Partmann *et al.* in prep), we present a new set of simulations that we use to explore the effect of the parameter choices of our feedback model in more detail as well as the stellar dynamics in the vicinity of the BH.

References

Birchall K. L., Watson M. G., & Aird J. 2020 *MNRAS*, 492, 2268

Greene J. E., Strader J., & Ho L. C. 2020 *ARAA*, 58, 257

Hu C.-Y., Naab T., Walch S., Moster B. P., & Oser L. 2014 *MNRAS*, 443, 1173

Hu C.-Y., Naab T., Walch S., Glover S. C. O., & Clark P. C. 2016 *MNRAS*, 458, 3528

Hu C.-Y., Naab T., Glover S. C. O., Walch S., & Clark P. C. 2017 *MNRAS*, 471, 2151

Koudmani S., Sijacki D., & Smith M. C. 2022 *MNRAS*, 516, 2112

Lahén N., Naab T., Kauffmann G., Szecsi D., Hislop J. M., et al. 2023 *MNRAS*, 522, 3092

Mezcua M., & Sánchez H. D. 2020 *ApJ*, 898, L30

Mezcua M., Siudek M., Suh H., Valiante R., Spinoso D., & Bonoli S. 2023 *ApJ*, 943, L5

Ostriker J. P., Choi E., Ciotti L., Novak G. S., & Proga D. 2010 *ApJ*, 722, 642

Partmann C., Naab T., Lahén N., Churazov E., Hislop J., Rantala A., et al. *in prep*

Rantala A., Pihajoki P., Johansson P. H., Naab T., Lahén N., & Sawala T. 2017 *ApJ*, 840, 53

Schutte Z., & Reines A. E. 2022 *Nature*, 516, 2112

Sharma R. S., Brooks A. M., Somerville R. S., Tremmel M., et al. 2020 *ApJ*, 897, 103

Silk J., 2017 *ApJ*, 839, L13

Springel, V. 2005 *MNRAS*, 364, 1105

Steinwandel U. P., Moster B. P., Naab T., Hu C.-Y., & Walch S. 2020 *MNRAS*, 495, 1035

Zaw I., Rosenthal M. J., Katkov I. Y., Gelfand J. D., Chen Y.-P., et al. 2023 *ApJ*, 897, 111

Black Hole Winds at All Scales
Proceedings IAU Symposium No. 378, 2024
G. Bruni, M. Diaz Trigo, S. Laha & K. Fukumura, eds.
doi:10.1017/S1743921323003344

Atomic and Molecular Gas Outflows in FIR Bright QSOs at High-z

Kirsty M. Butler[1,2]

[1]Institut de Radioastronomie Millimétrique (IRAM), 300 rue de la Piscine, 38400 Saint-Martin-d'Héres, France
email: butler@iram.fr

[2]Leiden Observatory, Leiden University, PO Box 9513, 2300 RA Leiden, the Netherlands

Abstract. Feedback and outflows associated with a quasar phase are expected to be critical in quenching the most massive galaxies at high-z. Observations targeting the cool molecular and atomic phases, which dominate the mass and momentum budget of massive galaxy outflows and remove the direct fuel for star formation are, however, severely limited in high-z QSO hosts. We discuss two recent ALMA programs: one targeting molecular outflows in 3 $z \sim 6$ QSO hosts using the OH 119 μm absorption line and another targeting the diffuse, predominantly atomic gas in the halos surrounding 5 QSO host between $z \sim 2-4$ using the $OH^+(1_1-1_0)$ absorption line. Outflows are successfully detected in both samples and compared with outflows driven by high-z star-forming galaxies observed in the same lines. Both studies indicate that observing QSOs during the blow-out phase is crucial for studying the impact of the active nucleus on the ejection of gas from the host galaxy.

Keywords. galaxies: high-redshift — galaxies: starburst — quasars: general

1. Introduction

Feedback from star formation and black hole activity plays a key role in the evolution of galaxies and galaxy populations embedded in the cosmic web. Galaxy evolution is regulated by energy and momentum injected into the ISM via these processes which heat, disturb or eject gas that may have fueled future star formation. Outflows are an essential phenomenon required by models and simulations to reproduce observed disky-galaxy morphologies, metallicity gradients and the polluted circum-/inter galactic medium (Governato *et al.* 2010; Veilleux *et al.* 2005; Simcoe *et al.* 2004) as they preferentially remove low angular momentum and metal-enriched material from the centres of galaxies.

Regulated galaxy growth implies that the most rapidly evolving galaxies must also experience the most vigorous feedback. Luminous AGNs, or QSOs, are believed to play a major role in quenching star formation in the most massive galaxies, primarily dispersing the available molecular gas through a blow–out phase (Costa *et al.* 2018), halting star formation and exposing the previously obscured QSO nucleus. Energy from the central AGN can then escape uninhibited, injecting heat and turbulence directly into the circumgalactic medium (CGM) and preventing this material from falling back in and re-fueling the galaxy. Assuming this evolutionary track, Dusty Star-Forming Galaxies (DSFGs) harbouring QSOs are currently in their transition phase, however, the fraction of star-forming galaxies identified as optically luminous is small. This suggests a fast ($\sim$ 1Myr; Simpson *et al.* 2012) and therefore explosive transition phase resulting in significant gas outflows.

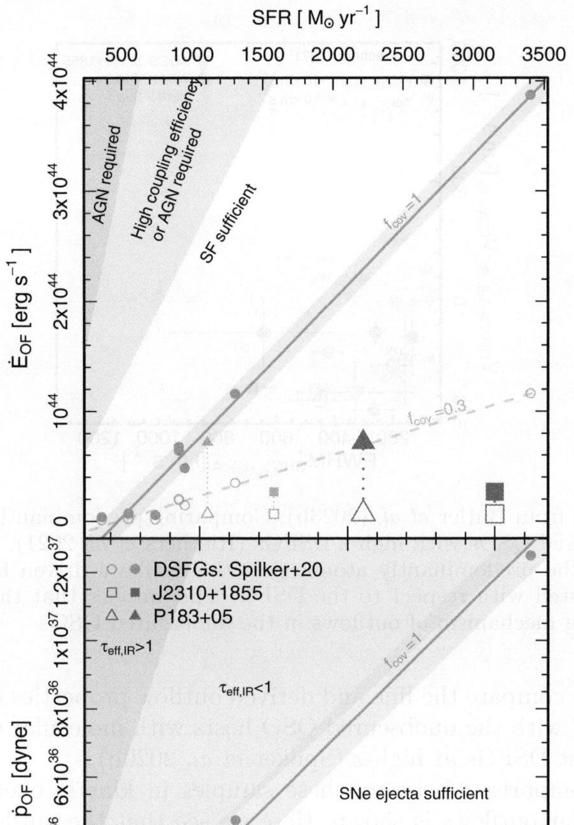

Figure 1. From Butler *et al.* (2023a): Comparing Molecular outflows from high-z unobscured QSOs with high-z DSFGs (Spilker *et al.* 2020b). The kinetic energy flux (top) and momentum flux (bottom) as a function of FIR luminosity and SFR. The unobscured QSOs (orange) do not require input from an AGN to drive their molecular outflows and in fact, appear to be suppressed after the blow-out phase compared to the DSFGs. This implies that the central AGN is not the dominant driving mechanism of outflows in the unobscured QSOs.

At high-z considerable work has begun on detecting and studying cool gas outflows in DSFGs using the $OH^+(1_1 - 1_0)$ (Shao *et al.* 2022; Riechers *et al.* 2021; Butler *et al.* 2021, 2023b) and OH 119 μm (Spilker *et al.* 2020b; Butler *et al.* 2023a) absorption lines, successfully showing that these lines are sensitive and reliable tracers. Observations in high-z QSOs, however, are greatly lacking. Here, we discuss two recent investigations of cool gas outflows in high-z QSO hosts using the $OH^+(1_1 - 1_0)$ and OH 119 μm absorption lines, aiming to fill this gap.

2. Overview

Butler *et al.* (2023a,c) presents OH 119 μm observations of three $z > 6$ unobscured dusty QSOs. They find blueshifted absorption, indicating molecular outflows, in two of

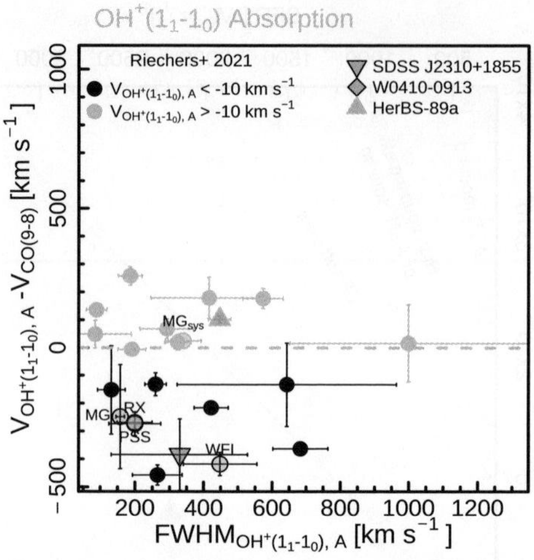

Figure 2. Adapted from Butler *et al.* (2023b): Comparing predominantly atomic gas outflows from high-z unobscured QSOs with high-z DSFGs (Riechers *et al.* 2021). The line width versus outflow velocity of the predominantly atomic gas: the outflows driven from the QSO sample (green) are not boosted with respect to the DSFG. This implies that the central AGN is not the dominant driving mechanism of outflows in the unobscured QSOs.

their sources. They compare the line and derived outflow properties of the two molecular outflows associated with the unobscured QSO hosts with molecular outflows observed in similarly FIR bright DSFGs at high-z (Spilker *et al.* 2020b).

In Fig. 1 the comparison between these samples in kinetic energy and momentum flux of the molecular outflows is shown. Here we see that the outflows driven from the unobscured QSO hosts are suppressed with respect to the DSFGs.

Butler *et al.* (2023b) targeted five high-z QSO hosts (four unobscured and one obscured) with the $OH^+(1_1 - 1_0)$ line. Outflows were detected in three QSO hosts with an additional tentative detection in a fourth and no detection of absorption in the fifth. The one obscured QSO host was the only source to display systemic OH^+ absorption (along with its blue-shifted absorption). Fig. 2 compares the full-width half maximum and Doppler shifted velocity of the OH^+ absorption components between the QSO and DSFG samples. They find no offset in either outflow velocity or outflow line width between the two samples, again providing no indication that the central AGN significantly contributes to or boosts the neutral atomic outflows in high-z QSOs.

3. Implications

Despite the dusty QSO hosts in these samples harbouring both an active nucleus and comparable star formation rates as their comparison DSFG samples, their neutral atomic and molecular outflows are not boosted. Both studies, however, note that they contain only unobscured QSOs (except one obscured QSO in Butler *et al.* 2023b), which have already gone through their 'blow-out' phase in which the obscuring material directly surrounding the central AGN is removed. During this phase, radiation from the AGN is trapped inside the QSO host and is believed to significantly contribute to the outflow of central gas via radiation pressure (Costa *et al.* 2018). Once removed, radiation is able to efficiently escape the host galaxy through the cleared path and thus may contribute significantly less to the ejection of cool gas.

Venemans *et al.* (2018, 2020) showed in their studies of high-z unobscured QSOs that no trend between the FIR luminosity, both spatially integrated and in just the central regions, and the UV magnitude of the central QSO is seen. Observation of high-J CO lines in several sources (Gallerani *et al.* 2014; Decarli *et al.* 2022) similarly provides a picture where the central QSO does not contribute significantly to the host galaxy. Thus, the conclusion of the studies discussed in this proceeding is that QSO hosts currently going through their blow-out phase should be the targets for studies aiming to investigate the true impact of the central active nucleus on the ejection of neutral and molecular gas from the host galaxy. In this way, we will be able to capture the point in a galaxy's evolution where the central active nucleus is most directly impactful on the host galaxy and therefore provide a better understanding of its role in the quenching of massive galaxies at high-z.

References

Butler, K. M., van der Werf, P. P., Rybak, M., et al. 2021, *ApJ*, 919, 5.
Butler, K. M., van der Werf, P. P., Topkaras, T., 2023, *ApJ*, 944, 134.
Butler, K. M., van der Werf, P. P., Topkaras, T., 2023, *ApJ*, 949, 122.
Butler, K. M., van der Werf, P. P., Omont, A., et al. 2023, *A&A*, 674, L5.
Costa, T., Rosdahl, J., Sijacki, D., & Haehnelt, M. G. 2018, *MNRAS*, 473, 4197
Decarli, R., Pensabene, A., Diax-Santos, T., et al. 2022, *A&A*, 662, A60
Spilker, J. S., Aravena, M., Phadke, K. A., et al. 2020, *ApJ*, 905, 86.
Gallerani, S., Ferrara, A., Neri, R., & Maiolino, R. 2014, *MNRAS*, 445, 2848
Governato, F., Brook, C., Mayer, L., et al. 2010, *Nature*, 463, 203
Riechers, D. A., Cooray, A., Pérez-Fournon, I., & Neri, R., *ApJ*, 913, 141.
Shao, Y., Wang, R., Weiss, A., et al. 2022, *A&A*, 668, A121
Simcoe, R. A., Sargent, W. L. W., & Rauch, M. 2004, *ApJ*, 1060 606, 92
Simpson, J. M., Smail, I., Swinbank, A. M., et al. 2012, *MNRAS*, 426, 3201
Veilleux, S., Cecil, G., & Bland-Hawthorn, J. 2005, *ARAA*, 43, 769,
Venemans, B. P., Decarli, R., Walter, F., et al. 2018, *ApJ*, 1092 866, 159
Venemans, B. P., Walter, F., Neeleman, M., et al. 2020, 1094 *ApJ*, 904, 130

Black Hole Winds at All Scales
Proceedings IAU Symposium No. 378, 2024
G. Bruni, M. Diaz Trigo, S. Laha & K. Fukumura, eds.
doi:10.1017/S1743921323003265

Super massive black hole and host galaxy growth at the onset of strong feedback at $z = 6 - 7.5$

Chiara Feruglio[1], Manuela Bischetti[2], Roberta Tripodi[1,2], Fabrizio Fiore[1], Enrico Piconcelli[3] and Luca Zappacosta[3]

[1]INAF - Osservatorio Astronomico di Trieste, Via G. Tiepolo 11, I-34143 Trieste, Italy
email: chiara.feruglio@inaf.it

[2]Dipartimento di Fisica, Università di Trieste, Sezione di Astronomia, Via G.B. Tiepolo 11, I-34131 Trieste, Italy

[3]INAF - Osservatorio Astronomico di Roma, Via Frascati 33, I-00040 Monte Porzio Catone, Italy

Abstract. The first quasars at the Reionisation Epoch, z∼6-7.5, probe the early stages of supermassive black holes and host galaxy assembly. In this paper, we present recent results, exploiting VLT, ALMA and NOEMA observations, that allow us to constrain the onset of strong black hole feedback, the dust properties and star formation rates in high redshift quasars with unprecedented accuracy. These results highlight the strategic importance of ALMA high frequency (i.e. Band 9 and 8) observations to obtain a reliable overview of the host galaxy and supermassive black hole growth out to the highest redshifts.

Keywords. Quasars, Supermassive black holes, AGN host galaxies, Broad-absorption line quasar, High-redshift galaxies, Galaxy evolution

1. Introduction

Luminous quasars, powered by accretion onto supermassive black holes (SMBHs), already exist at the Epoch of Reionisation (EoR), when the Universe was only 0.5-1 Gyr old (Fan *et al.* 2022). Their BH masses are as large as those of luminous quasars at lower redshift, $M_{BH} = 10^9 - 10^{10}\ M_\odot$, meaning that BH growth had to be a fast process, and had to stop with a similarly high efficiency after the rapid build-up (Volonteri & Begelman 2010; Johnson & Haardt 2016). These huge BHs are overmassive compared to their host galaxies, with respect to the local BH mass - galaxy mass relation, suggesting that BH growth is dominating over galaxy growth in these systems.

Some feedback mechanism is then needed to slow down and suppress BH growth, leading to the symbiotic evolution observed in the local Universe. Candidate processes are inefficient gas accretion and/or feedback through BH winds (van der Vlugt & Costa 2019). At the same time, the host galaxies of high-z quasars are likely growing rapidly, which is also suggested by the recent JWST discovery of an excess of galaxies compared to the UV luminosity function predictions (Adams *et al.* 2023; Castellano *et al.* 2022; Finkelstein & Bagley 2022). The onset of significant BH feedback hampering BH growth would mark the transition from BH dominance to BH-and-galaxy symbiotic growth phases. Cosmological, hydrodynamic simulations of early BH and galaxy evolution support this scenario by identifying z∼6-7 as the transition epoch during which

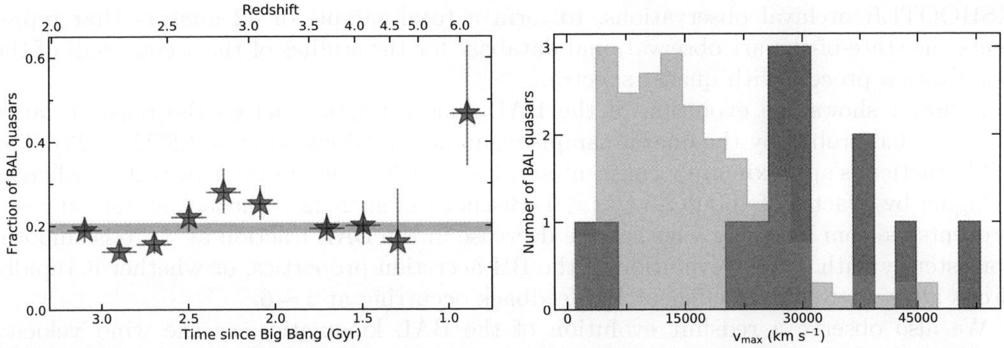

Figure 1. Left panel: Fraction of BAL quasars as a function of cosmic time, from 0.8 to 3.2 Gyr, in bins of ~ 200 Myr, based on the quasar sample in Bischetti *et al.* 2023. The shaded region corresponds to the BAL fraction measured at $2.1 < z < 4.6$ in Bischetti *et al.* 2022. Top axis indicates the corresponding redshift. Right panel: the BAL velocity distribution, parametrised here with the maximum BAL velocity, v_{max}, for the SDSS sample (blue histogram) and the XQR-30 sample (grey histogram).

quasar feedback increases in strength and starts to significantly slow down BH growth (Costa *et al.* 2018; Hu *et al.* 2022; Ishibashi & Fabian 2023).

To put this scenario into a quantitative observational framework it is necessary to accurately measure galaxy masses and growth rates and compare them to the BH growth rates, so to derive the mean evolutionary paths of quasars in the $M_{\rm BH} - M_{\rm dyn}$ plane, and verify where the BH-dominance curve changes slope, thus leading to the SMBH-host galaxy symbiotic growth. Luminous quasars at EoR may be the tip of the iceberg of a large AGN population, but allow the investigation of the BH accretion process and its feedback on the host galaxy at its extreme, and in the cleanest way (Fiore *et al.* 2017; Bischetti *et al.* 2017). The recent JWST discovery of many AGN at high redshift suggests that the AGN fraction at high z is larger than previously thought (Übler *et al.* 2023), showing that BH dominance may not only be in place for BHs with $M_{\rm BH} > 10^9$ M$_\odot$, but also for less massive ones, and suggesting that luminous quasars likely represent the bright end of a population of smaller systems in which the BH grows faster than the host galaxy (Harikane *et al.* 2023; Kocevski *et al.* 2023).

In this paper we present an overview on recent results on strong quasar feedback (Sec. 2), and on the concurrent growth of SMBH and their host galaxies in quasars at $z > 6$ (Sec. 3).

We adopt a ΛCDM cosmology from the Planck Collaboration (2018): $H_0 = 67.4$ km s^{-1} Mpc^{-1}, $\Omega_m = 0.315$ and $\Omega_\Lambda = 0.685$.

2. The onset of strong quasar feedback at the Reionisation Epoch

Bischetti *et al.* (2022) analysed quasars from the X-shooter legacy survey of Quasars at Reionisation (D'Odorico *et al.* 2023), to investigate the prevalence and kinematics of BH-driven outflows as traced by Broad Absorption Line (BAL) features at $z > 6$. They analysed also 1905 luminous (bolometric luminosity $L_{\rm bol} \gtrsim 10^{46.5}$ erg s^{-1}) quasars at $z = 2.1 - 5$, drawn from the Sloan Digital Sky Survey (Shen *et al.* 2011), using the same methodology as for the the high-redshift XQR-30 sample, to derive the evolution of the BAL fraction with cosmic time. XQR-30 is a sample of 30 spectra of quasars probing the EoR ($5.8 \leq z \leq 6.6$) observed at high resolution ($R \sim 10,000$) and high signal-to-noise ratio (SNR $\sim 11 - 41$ per bin of 10 km s^{-1}) within an ESO XSHOOTER Large Programme of 248 h. The XQR-30 observations were complemented with all the available

XSHOOTER archival observations, to form a total sample of 42 quasars that represents the state-of-the-art observational database for the studies of the second-half of the reionization process with quasar spectra.

Figure 1 shows the evolution of the BAL quasar fraction across the redshift range $2.1 < z < 6.6$ probed by the quasar sample examined in Bischetti *et al.* (2022, 2023). The BAL fraction is approximately constant at $z \lesssim 4.5$ with a median value of $\sim 20\%$, whereas is higher by a factor of about 2.5 at $z \gtrsim 6$ (Bischetti *et al.* 2022). The lack of data at $z \sim 5$ prevents us from assessing whether the decrease in the BAL fraction at $z < 6$ is smooth, consistently with secular evolution of the BH accretion properties, or whether it rapidly drops after $z \sim 6$, due to efficient BH feedback occurring at $z \sim 6$.

We also observe a redshift evolution of the BAL kinematics, as the wind velocity increases at $z \gtrsim 4$. The typical BAL velocities at $z \sim 6$ are indeed a factor of 2-3 higher than those at $z < 4$ (Figure 1, right panel). These trends are consistent with BALs being more easily accelerated at early cosmic epochs, perhaps owing to a dustier or denser galaxy medium (Bischetti *et al.* 2023). By investigating the dependence of the BAL kinematics on the quasar luminosity and accretion rate, we were able to exclude that the redshift evolution is due to different quasar prperties in the sample.

These more common and faster BAL outflows at $z \sim 6$ may witness the rise of the feedback age, owing to the injection of large amounts of energy in the host galaxy interstellar medium (ISM), in agreement with expectations from galaxy evolution models (van der Vlugt & Costa 2019; Inayoshi *et al.* 2022). We conclude that BAL winds appear fundamental tools to assess importance of BH feedback at high redshift. However, this method also suffers from an intrinsic limitation, because the broad and often saturated BAL troughs associated with CIV prevent us from obtaining an accurate measurement of the outflowing gas mass, and hence of the energy injected into the galaxy ISM (Dunn *et al.* 2012; Borguet *et al.* 2013). To derive a rough estimate of the energy involved in BAL winds, we may assume that outflow masses at $z \sim 6$ are similar to those measured using unsaturated absorption lines in $z \sim 2$ BAL quasars (Moe *et al.* 2009; Dunn *et al.* 2010). Under this assumption, we find that BAL quasars at $z \sim 6$ globally inject about 20 times more energy with respect to $z \sim 2 - 4$ quasars (Bischetti *et al.* 2022) in the surrounding ISM, suggesting a phase of efficient BH feedback occurring at $z \sim 6$. The energy injected by these BAL outflows will likely suppress gas accretion and slow down BH growth (Torrey *et al.* 2020). This result represents the first observational evidence of an evolution that has so far only been predicted by cosmological simulations of the early assembly of bright quasars, in which SMBHs in bright quasars grow with accretion rates close to the Eddington limit or beyond at very high redshift ($z >> 6$, (Inayoshi *et al.* 2016; Pezzulli *et al.* 2016), while around $z = 6$ BH growth significantly slows down owing to the rise of strong feedback.

Evidences of strong feedback in these high redshift quasars come also from observations of the cold gas component, for which it is easier to derive the mass and energy rate properties. Figure 2 shows the [CII] 158 μm emission line from the cold ISM of quasar SDSS J2310+1855 (Tripodi *et al.* 2022). In this quasar, the high-velocity blue- and redshifted emission that is seen in the emission line profile are spatially located in the nuclear region, and do not follow the rotation curve of the main disk. The high excess in LOS velocity between these components and the best-fit disk ($v_{\rm LOS,disk} \sim$ 100 km s^{-1}) suggests that these blue- and red wings are due to a cold gaseous outflow, for which we estimate a outflow rate in the range $\dot{M}_{\rm out} \approx 1800 - 4500$ M$_\odot$ yr^{-1}, a kinetic power of $\dot{E}_{\rm out} = \frac{1}{2}\dot{M}_{\rm out} \times v_{\rm out}^2 = (1.5 - 3.7) \times 10^{44}$erg s^{-1}, and a wind momentum load $\frac{\dot{P}_{\rm out}}{\dot{P}_{\rm AGN}} = \frac{\dot{M}_{\rm out} \times v_{\rm out}}{L_{\rm bol}/c} = 0.6 - 1.4$, where we adopted a bolometric luminosity of

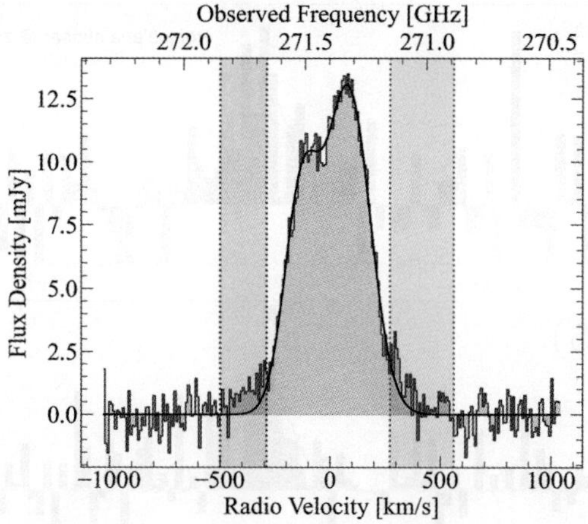

Figure 2. The [CII] 158 μm emission line profile of quasar SDSS J2310+1855 at $z = 6.0$ shows excess emission with respect to the gaussian best fit at the red and blue sides of the line. This excess emission is interpreted as evidence of a fast galactic outflow of dense molecular material (Tripodi *et al.* 2022).

$L_{\rm bol} = 3.13 \times 10^{47}$erg s^{-1}, derived from the rest-frame continuum at 3000 Å Bischetti *et al.* (2022) and the bolometric correction from Runnoe *et al.* (2012).

It is worth noting that this quasar hosts a ionised wind traced by a C IV BAL system with velocity $v_{\rm BAL} = 26900$ km s^{-1} (Bischetti *et al.* 2022), whose energy cannot be quantified due to the arguments discussed above, and a dense molecular outflow seen in absorption through blue-shifted (P-Cygni) profile in both OH 199μm and OH$^+(1_1 - 0_1)$ molecular transitions, probing cold and dense molecular material (Butler *et al.* 2023; Shao *et al.* 2022). These results suggest that outflows in these high redshift luminous quasars are truly multiphase. Further observations of molecular and ionic transitions with ALMA and NOEMA have the potential to assess whether there is a one-to-one correspondence between ionised winds traced by BAL systems and galaxy-scale outflows probed by molecular an ionic species in the far-infrared.

3. The host galaxy growth in $z > 6$ quasars

The host galaxy growth rate may be approximated by the ratio of the star formation rate to the total mass of the system, SFR/M$_{\rm host}$, where $M_{\rm host} = M_{\rm gas} + M_*$. We use $M_{\rm g+s}$, instead of $M_{\rm dyn}$, since the BH mass may not be negligible in some quasars hosting very massive BH. For example, quasar SDSS J0100+2802 has a very heavy BH, with $M_{\rm BH} \sim 40\% M_{\rm dyn}$ (Wang *et al.* 2019). Directly measuring the stellar mass M_* from observations has been out of reach so far for high redshift quasars, and can only be approximately evaluated by subtraction $M_{\rm dyn} - M_{\rm BH} - M_{\rm gas}$. A primary probe of molecular reservoirs, which dominate the $M_{\rm gas}$ term, are the carbon monoxide (CO) rotational transitions. At $z \sim 6 - 7.5$ the CO J=6-5 and 7-6 are conveniently redshifted at ~ 3 mm, and therefore are easily observable using ALMA and NOEMA. A recent, record-breaking CO detection was provided in the quasar J100758.264+211529.207 (Pōniuā'ena) at $z = 7.5149$ by Feruglio *et al.* (2023). Figure 3 shows the CO(6-5) and (7-6) emission line spectra from the host galaxy of the quasar obtained with NOEMA. The line luminosities correspond to a molecular mass $\mathrm{M(H_2)} = \alpha_{\rm CO} \times \mathrm{L'CO}(6-5)/\mathrm{r61} = (2.2 \pm 0.2) \times 10^{10}\ \mathrm{M_\odot}$, once a conversion factor $\alpha_{\rm CO} = 0.8\ \mathrm{M_\odot}\ (\mathrm{K\,km\,s^{-1}\,pc^2})^{-1}$ and a a line luminosity ratio $\mathrm{r61} = \mathrm{L'CO}(6-5)/\mathrm{L'CO}(1-0) = 0.75$ are adopted (Carilli *et al.* 2013). These results show that CO line detection are today feasible up to the highest redshifts. When CO

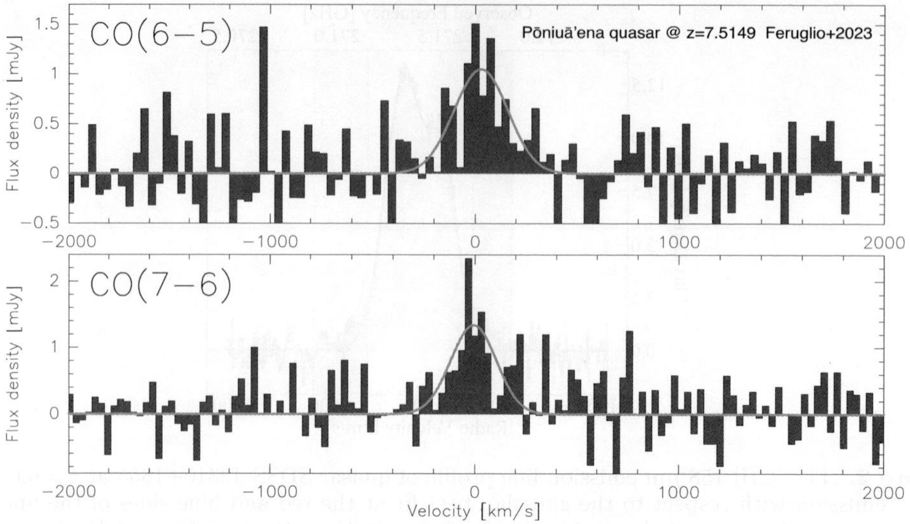

Figure 3. The CO(6-5) (upper panel) and (7-6) (lower panel) emission line spectra of the $z = 7.5149$ quasar Pōniuā'ena obtained with NOEMA Feruglio *et al.* 2023. The velocity zero point is set to the frequency 81.207 GHz for CO(6-5) and 94.734 GHz for CO(7-6). Magenta lines show the fit with a single Gaussian component.

lines are not available, one can approximate $M_{\rm host}$ with $M_{\rm dyn}$, which is routinely derived from $\sim$ kpc resolution observations using the bright [C II] line.

The SFR has proven, until recently, to be more difficult to measure. SFR in high-z quasars is typically derived by modelling the thermal dust emission probed by the (sub)-mm continuum. However, most of the values of SFR derived so far are poorly constrained because the emission peak of cold dust, located at $\lambda < 70\mu$m in the rest frame ($\nu > 4500$ GHz, i.e. > 600 GHz at $z \sim 6$) is not included in the fit (Wang *et al.* 2019; Novak *et al.* 2019), Fig. 4). This is why observations at > 600 GHz are essential to achieve tightly constrained SFR at this redshift. The commonly-used methodology is based on the modelling of the thermal dust emission with the following greybody distribution with a single temperature $T_{\rm dust}$:

$$S^{\rm obs}_{\nu/(1+z)} = \frac{\Omega}{(1+z)^3}[B_\nu(T_{\rm dust}(z)) - B_\nu(T_{\rm CMB}(z))](1 - e^{\tau_\nu}) \quad (1)$$

where $T_{\rm dust}(z) = (T_{\rm dust}^{(4+\beta)} + T_0^{(4+\beta)}[(1+z)^{(4+\beta)} - 1])^{\frac{1}{4+\beta}}$, $\tau_\nu = M_{\rm dust}k_\nu/A_{\rm gal}$ is the dust optical depth, $k_\nu = 0.45 \times (\nu/250{\rm GHz})^\beta\ cm^2\ g^{-1}$ is the dust opacity coefficient, and $A_{\rm gal}$ the galaxy area. Free parameters are the dust temperature $T_{\rm dust}$, the dust mass $M_{\rm dust}$ and the dust emissivity parameter β. Using the data typically available so far in the sub-mm range, $T_{\rm dust}$ remains unconstrained for most quasars resulting into SFR uncertain by at least a factor of ~ 5 (Wang *et al.* 2019). When only 1 or 2 data points at < 300 GHz are available, $M_{\rm dust}$ and SFR are uncertain by one order of magnitude. Similarly, assuming a fixed β leads to an estimate of $\rm M_{dust}$ with up to an order of magnitude uncertainty, and large systematic errors on important derived quantities, such as the gas-to-dust ratio and the molecular mass - thus biasing our understanding of the growth of these massive galaxies. Exceptions are the very few quasars detected by Herschel, which observed the peak of the cold dust SED emission, despite having coarse angular resolution.

ALMA Band 9 observations, in synergy with measurements at lower frequencies, represent the only method to measure reliably $\rm T_{dust}$ and SFR (Tripodi *et al.* 2023). Fig. 4 shows the first accurate $\rm T_{dust}$ determination in a $z > 6$ quasar, $\rm T_{dust} = 40 \pm 2$ K.

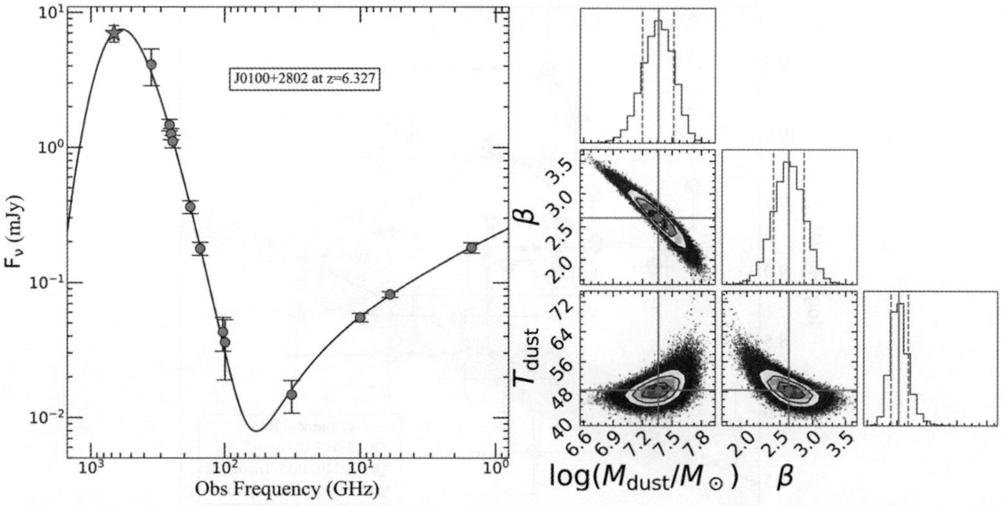

Figure 4. Left panel: Spectral Energy Distribution of quasar J0100+2802 using ALMA 670.91 GHz data (filled star, Tripodi *et al.* (2023)), the continuum fluxes from 32 GHz to 353 GHz (Wang *et al.* 2019), and at 1.5, 6 and 10 GHz (Li *et al.* 2022) (filled circles). The best-fitting curve is shown by the solid line. Right panel: posterior probability distributions of $T_{\rm dust}$, $M_{\rm dust}$, β. Solid lines indicate the best-fitting value for each parameter. Dashed lines mark the 16th and 84th percentiles for each parameter.

We estimate the total infrared (TIR) luminosity for the best-fit model by integrating from 8 to 1000 μm rest-frame, obtaining $L_{\rm TIR} = 5.30 \pm 0.64 \times 10^{12}$ L$_\odot$. Observations and radiative transfer simulations suggest that the radiative output of luminous QSOs substantially contributes to dust heating on kpc scale (Schneider *et al.* 2015; Di Mascia *et al.* 2021; Walter *et al.* 2022). Duras *et al.* (2017) showed that on average $\sim 50\%$ of the total IR luminosity in QSOs with $L_{\rm bol} > 10^{47}$ erg s^{-1} is due to dust heated by the AGN radiation. Applying this prescription, we obtain SFR $= 265 \pm 32$ M$_\odot$yr^{-1} for the quasar J0100+2802. These observations allowed us to derive SFR with 25% accuracy. This would not have been possible without a Band 9 observation with ALMA.

A further, important complication to be considered is represented by steep temperature gradients due to the AGN contribution to the dust heating, that are predicted by simulations (Di Mascia *et al.* 2023) and have been detected in quasars at $z = 4.4$ (Tsukui *et al.* 2023) and $z = 6.9$ (Walter *et al.* 2022). The latter suggests a higher T$_{\rm dust}$ in the center (> 130 K) while the overall SED is dominated by a colder component (T$_{\rm dust}$ $\sim$50 K). These gradients advocate for spatially resolved T$_{\rm dust}$ maps to account for AGN heating, which are also feasible today with ALMA at least for the brightest quasars at $z > 6$.

Together these ingredients allow us to evaluate the evolutionary state of the SMBH - host galaxy system. Figure 5 shows $M_{\rm BH}$ versus $M_{\rm dyn}$ for J0100+2802, J2310+1855 (Tripodi *et al.* 2022) and a compilation of QSOs at different redshifts. The majority of QSOs, including J0100+2802, lie above the local relation in the BH dominance regime (Volonteri 2012). We note that this tension can be partially softened if accounting for the uncertainties on the dynamical mass estimate, that mainly depend on the determination of the disk inclination and can be significantly high for some QSOs (Valiante *et al.* 2014; Pensabene *et al.* 2020). For quasar J0100+2802, $M_{\rm dyn}$ can be as high as 10^{12} M$_\odot$ if we account for this uncertainty. For J0100+2802, Tripodi *et al.* (2023) find $(1-\epsilon)\dot{M}_{\rm BH}/M_{\rm BH} > {\rm SFR}/M_{\rm g+s}$, suggesting that the BH is dominating the BH-galaxy growth. We note that their results do not consider the gas inflow, which is currently challenging to determine from observations. However, this term is expected to

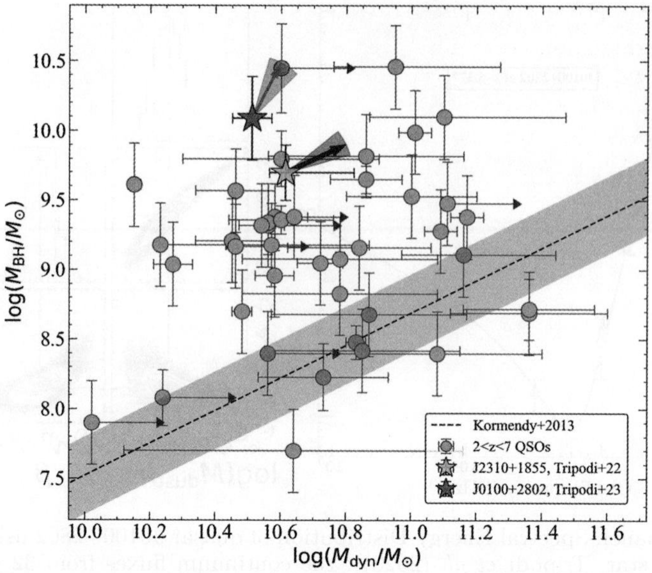

Figure 5. BH mass vs dynamical mass for J0100+28 (magenta symbol), compared with QSO J2310+18 at $z = 6.0028$ (blue symbol), and a compilation of $z \sim 2-7$ quasars from Tripodi *et al.* (2023) and references therein (grey symbols). The black dashed line (and shaded area) is the local relation from Kormendy & Ho (2013). For J0100+28 (J2310+18), the slope of the arrow, with its uncertainty, indicates the growth efficiency of the SMBH with respect to the growth of the host galaxy.

contribute to both SFR and M_{g+s}, leaving their ratio mostly unaffected. Conversely, in the quasar J2310+1855 at $z \sim 6$ AGN feedback might be slowing down the accretion onto the SMBH, while the host galaxy grows fast (Tripodi *et al.* 2022; Bischetti *et al.* 2022). The different evolutionary state of J0100+2802 and J2310+1855, which are separated by only ~ 60 Myr (i.e. $\Delta z \sim 0.3$), arise mainly from the difference in their BH mass ($M_{BH,J0100+28} \sim 2 \times M_{BH,J2310+18}$) and SFR ($SFR_{J0100+28} \sim 0.2 \times SFR_{J2310+18}$). As already mentioned, SFR/M_* is a better probe of the galaxy growth, however this is not available for most high redshift QSOs. For J0100+2802, using SFR/M_* would not affect our results since $M_{g+s} \approx M_*$. For J2310+1855, $M_* \ll M_{g+s}$, therefore it would imply an even flatter slope.

The cases of these two quasars represent two snapshots of different evolutionary phases, and are useful to illustrate the methodology that could be used to set this evolutionary scenario in a quantitative observational framework. This goal can be achieved by using a large sample of quasars at $z > 6$ and accurately measure their BH masses, dynamical masses and SFR, so to derive the mean evolutionary paths of quasars in the $M_{BH} - M_{dyn}$ plane.

References

Adams, N. J., Conselice, C. J., Austin, D., et al. 2023, *arXiv*, doi:10.48550/arXiv.2304.13721
Bischetti, M., Piconcelli, E., Vietri, G., et al. 2017, *A&A*, 598, A122.
Bischetti, M. and Feruglio, C. and D'Odorico, V. et al. 2022, *Nature*, 605, 7909
Bischetti, M., Fiore, F., Feruglio, C., et al. 2023, *arXiv*, doi:10.48550/arXiv.2301.09731
Borguet, B. C. J., Arav, N., Edmonds, D., et al. 2013, *ApJ*, 762, 49
Butler, K. M., van der Werf, P. P., Topkaras, T., et al. 2023, *ApJ*, 944, 134
Castellano, M., Fontana, A., Treu, T., et al. 2022, *ApJL*, 938, L15
Costa, T., Rosdahl, J., Sijacki, D., et al. 2018, *MNRAS*, 479, 2079

Di Mascia, F., Gallerani, S., Behrens, C., et al. 2021, *MNRAS*, 503, 2349
Di Mascia, F., Carniani, S., Gallerani, S., et al. 2023, *MNRAS*, 518, 3667
D'Odorico, V., Bañados, E., Becker, G. D., et al. 2023, *MNRAS*, 523, 1399
Draine, B.T. 2003, *ARAA*, 41, 241
Dunn, J. P., Bautista, M., Arav, N., et al. 2010, *ApJ*, 709, 611
Dunn, J. P., Arav, N., Aoki, K., et al. 2012, *ApJ*, 750, 143
Duras, F., Bongiorno, A., Piconcelli, E., et al. 2017, *A&A*, 604, A67
Fan, X., Banados, E., & Simcoe, R. A. 2022, *arXiv*, doi:10.48550/arXiv.2212.06907
Finkelstein, S. L. & Bagley, M. B. 2022, *ApJ*, 938, 25
Fiore, F., Feruglio, C., Shankar, F., et al. 2017, *A&A*, 601, A143.
Harikane, Y., Zhang, Y., Nakajima, K., et al. 2023, *arXiv*, doi:10.48550/arXiv.2303.11946
Hu, H., Inayoshi, K., Haiman, Z., et al. 2022, *ApJ*, 934, 132
Inayoshi, K., Haiman, Z., & Ostriker, J. P. 2016, *MNRAS*, 459, 3738
Inayoshi, K., Nakatani, R., Toyouchi, D., et al. 2022, *ApJ*, 927, 237
Ishibashi, W. & Fabian, A. C. 2023, *MNRAS*, 519, 1931
Johnson, J. L. & Haardt, F. 2016, *PASA*, 33, e007
Kocevski, D. D., Onoue, M., Inayoshi, K., et al. 2023, *arXiv*, doi:10.48550/arXiv.2302.00012
Kormendy, J. & Ho, L. C. 2013, *ARAA*, 51, 511
Moe, M., Arav, N., Bautista, M. A., et al. 2009, *ApJ*, 706, 525
Novak, M., Bañados, E., Decarli, R., et al. 2019, *ApJ*, 881, 63
Pensabene, A., Carniani, S., Perna, M., et al. 2020, *A&A*, 637, A84
Pezzulli, E., Valiante, R., & Schneider, R. 2016, *MNRAS*, 458, 3047
Planck Collaboration et al. 2018, *arXiv*, 1807.06209
Runnoe, J. C., Brotherton, M. S., & Shang, Z. 2012, *MNRAS*, 422, 478
Schneider, R., Bianchi, S., Valiante, R., et al. 2015, *A&A*, 579, A60
Shao, Y., Wang, R., Weiss, A., et al. 2022, *A&A*, 668, A121
Shen, Y., Richards, G. T., Strauss, M. A., et al. 2011, *ApJS*, 194, 45
Torrey, P., Hopkins, P. F., Faucher-Giguère, C.-A., et al. 2020, *MNRAS*, 497, 5292
Tripodi, R., Feruglio, C., Fiore, F., et al. 2022, *A&A*, 665, A107
Tripodi, R., Feruglio, C., Kemper, F., et al. 2023, *ApJL*, 946, L45
Tsukui, T., Wisnioski, E., Krumholz, M. R., et al. 2023, *arXiv*, doi:10.48550/arXiv.2302.07272
Übler, H., Maiolino, R., Curtis-Lake, E., et al. 2023, *arXiv*, doi:10.48550/arXiv.2302.06647
Valiante, R., Schneider, R., Salvadori, S., et al. 2014, *MNRAS*, 444, 2442
van der Vlugt, D. & Costa, T. 2019, *MNRAS*, 490, 4918
Volonteri, M. & Begelman, M. C. 2010, *MNRAS*, 409, 1022
Volonteri, M. 2012, *Science*, 337, 544.
Walter, F., Neeleman, M., Decarli, R., et al. 2022, *ApJ*, 927, 21
Wang, F., Wang, R., Fan, X., et al. 2019, *ApJ*, 880, 2

Chapter 4:
Outflows from stellar black holes

Black Hole Winds at All Scales
Proceedings IAU Symposium No. 378, 2024
G. Bruni, M. Diaz Trigo, S. Laha & K. Fukumura, eds.
doi:10.1017/S1743921324000437

The role of outflows in black-hole X-ray binaries

Nikolaos D. Kylafis[1,2] **and Pablo Reig**[2,1]

[1]University of Crete, Physics Department, 70013 Heraklion, Greece
email: kylafis@physics.uoc.gr

[2]Institute of Astrophysics, FORTH, 70013 Heraklion, Greece
email: pau@physics.uoc.gr

Abstract. An outflow, from the hot inner flow, in black-hole X-ray binaries is always expected due to the positive Bernoulli integral in the hot inner flow. We have demonstrated that, if one considers this outflow as the place where not only Comptonization occurs, but also radio emission, many observed correlations, including the long-standing one between radio and X-rays, can be explained with one simple model.

Keywords. X-ray binaries, black holes, outflows, jets

1. Introduction

Black-hole X-ray binaries (BXHRBs) exhibit many observational correlations, the most prominent being the correlation seen in the hard state between the radio flux F_R, say at 8.6 GHz, and the X-ray flux F_X, say between 2 and 10 keV (Hannikainen et al. 1998; Corbel et al. 2000, 2003; Gallo et al. 2003; Bright et al. 2020; Shaw et al. 2021). In the so-called "standard track", the correlation is of the form $F_R \propto F_X^{\beta}$, where $\beta \approx 0.7$ (Gallo et al. 2003; Corbel et al. 2003)) or $\beta \approx 0.6$ (Corbel et al. 2013). Radio and X-ray emission occurs also in the hard-intermediate state, but the relation between the two has not been studied yet.

The standard picture of the accretion flow in BHXRBs in their hard and hard-intermediate states is that it consists of a geometrically thin, optically thick, outer disk (Shakura & Sunyaev 1973)) and a hot inner flow, which is geometrically thick and optically thin (Narayan & Yi 1994, 1995; Abramowicz et al. 1995). This hot inner flow is taken as the corona in many models. It is, unfortunately, not widely recognized that, due to the positive Bernoulli integral of the hot inner flow, a mildly relativistic outflow *must* emanate from it (Blandford & Begelman 1999; Kazanas 2015). In other words, *the Comptonizing corona is not static, but an outflowing one.* This has tremendous implications.

Contrary to the mainstream, where people use a static corona, few people examine an outflowing corona (see, however, Beloborodov 1999). An important exception is the Jet Emitting Disk/Standard Accretion Disk (JED/SAD) model (Marcel et al. 2019; Barnier et al. 2002). We have been promoting the idea that Comptonization in BHXRBs takes place mainly in the outflow (outflowing corona). Our picture explains naturally a number of correlations, some of which have not been explained by any other model. Due to the magnetic field that is needed for the ejection of the outflow (e.g., Blandford & Payne 1982), the outflow emits also radio waves. Since it is the same electrons that do the Compton upscattering of the soft photons and the radio emission by synchrotron, it is not surprising that the radio and the X-rays are correlated.

For the radio – X-ray correlation, we restrict ourselves to GX 339-4, because it is well studied and we have modeled various correlations exhibited by this source (Reig & Kylafis 2015; Kylafis & Reig 2018; Kylafis, Reig, & Papadakis 2020).

2. The model

Guided by the observations, our model consists of a *parabolic* outflow with two symmetric lobes, where both Comptonization and radio emission occur. Soft photons from the accretion disk enter the outflow at its bottom and inverse Compton scattering produces the observed power law in hard X-rays. The same electrons that do the Compton up-scattering, also emit radio waves. Details of the model can be found in our work over the last 20 years (Reig, Kylafis, & Giannios 2003; Giannios, Kylafis, & Psaltis 2004; Giannios 2005; Kylafis et al. 2008; Reig & Kylafis 2015; Kylafis & Reig 2018; Reig et al. 2018; Reig & Kylafis 2019; Kylafis, Reig, & Papadakis 2020; Reig & Kylafis 2021; Kylafis, Reig, & Tsouros 2023, submitted to A&A).

3. Results

Despite its simplicity, our model explains quantitatively many correlations and observations, some of which have not been explained by any other model. These are:

1) the energy spectrum from radio to hard X-rays for the source XTE J1118+480 (Giannios 2005),

2) the $F_R - F_X$ correlation in GX 339-4 (Kylafis, Reig, & Tsouros 2023, A&A, submitted),

3) the time-lag – Fourier frequency correlation in Cyg X-1 (Reig, Kylafis, & Giannios 2003),

4) the correlation between the time-lag and the photon-number spectral index Γ in GX 339-4 (Kylafis & Reig 2018) and other sources (Reig et al. 2018),

5) the fact that this correlation depends on the inclination of the source (Reig & Kylafis 2019),

6) the phase-lag – cutoff-energy correlation observed in GX 339-4 (Reig & Kylafis 2015),

7) the narrowing of the auto-correlation function with increasing photon energy seen in Cyg X-1 (Giannios, Kylafis, & Psaltis 2004),

8) the correlation between the Lorentzian frequencies in the power spectrum and the photon-number spectral index Γ in Cyg X-1 and GX 339-4 (Kylafis et al. 2008),

9) the photon-number spectral index as a function of phase of the type-B QPO in GX 339-4 (Kylafis, Reig, & Papadakis 2020),

10) the outflow provides a natural *lamppost* for the hard X-ray photons that return to the disk, where reflection and reverberation occurs (Reig & Kylafis 2021), and

11) last but not least, the outflow model may explain naturally the observed X-ray polarization in BHXRBs, since most of the matter of the outflow is at its bottom, i.e., as a "slab", and a slab is invoked to explain the polarization in BHXRBs (Krawczynski et al. 2022; Kylafis, Reig, & Tsouros 2023, A&A, submitted).

All of the above correlations are explained with only two parameters of the outflow: the radius R_0 at its base and the Thomson optical depth $\tau_{||}$ along its axis. For all the observations and correlations above, these two quantities vary in our model in the same narrow ranges $10 \lesssim R_0/R_g \lesssim 1000$ and $1 \lesssim \tau_{||} \lesssim 10$.

4. Conclusions

According to the statistician George E. P. Box, "All models are wrong, some are useful". Due to its simplicity and its various successes, our model might be useful.

References

Abramowicz, M. A., Chen, X., Kato, S., & Regev, O. 1995, ApJ, 438, L37
Barnier, S., Petrucci, P.-O., Ferreira, J., et al. 2022, A&A, 657, A11
Beloborodov, A. M. 1999, ApJ, 510, L123
Blandford, R. D., & Begelman, M. C. 1999, MNRAS, 303, L1
Blandford, R. D., & Payne, D. G. 1982, MNRAS, 199, 883
Bright, J. S. Fender, R. P., Motta, S. E., et al. 2020, NatAs, 4, 697
Corbel, S., Coriat, M., Brocksopp, C., et al. 2013, MNRAS, 428, 2500
Corbel, S., Fender, R. P., Tzioumis, A. K., Nowak, M., McIntyre, V., Durouchoux, P., & Sood, R. 2000, A&A, 359, 251
Corbel, S., Nowak, M. A., Fender, R. P., Tzioumis, A. K., & Markoff, S. 2003, A&A, 400, 1007
Gallo, E., Fender, R. P., & Pooley, G. G. 2003, MNRAS, 344, 60
Giannios, D. 2005, A&A, 437, 1007
Giannios, D., Kylafis, N. D., & Psaltis, D. 2004, A&A, 425, 163
Hannikainen, D. C., Hunstead, R. W., Campbell-Wilson, D., & Sood, R. K. 1998, A&A, 337, 460
Kazanas, D. 2015, ASSL, 414, 207
Krawczynski, H, Muleri, F., Dovciak, M., et al. 2022, Science, 378, 650
Kylafis, N. D., Papadakis, I. E., Reig, P., Giannios, D., & Pooley, G. G. 2008, A&A, 489, 481
Kylafis, N. D., & Reig, P. 2018, A&A, 614, L5
Kylafis, N. D., Reig, P., & Papadakis, I. 2020, A&A, 640, 16
Marcel, G., Ferreira, J., Clavel, M., et al. 2019, A&A, 626, A115
Narayan, R., & Yi, I. 1994, ApJ, 428, L13
Narayan, R., & Yi, I. 1995, ApJ, 452, 710
Reig, P., & Kylafis, N. D. 2015, A&A, 584, 109
Reig, P., & Kylafis, N. D. 2019, A&A, 625, 90
Reig, P., & Kylafis, N. D. 2021, A&A, 646, 112
Reig, P., Kylafis, N. D., and Giannios, D. 2003, A&A, 403, L15
Reig, P., Kylafis, N. D., Papadakis, I. E. & Costado, M. 2018, MNRAS, 473, 4644
Shakura, N. I., & Sunyaev, R. A. 1973, A&A, 24, 337
Shaw, A. W., Plotkin, R. M., Miller-Jones, J. C. A., et al. 2021. ApJ, 907, 34

Black Hole Winds at All Scales
Proceedings IAU Symposium No. 378, 2024
G. Bruni, M. Diaz Trigo, S. Laha & K. Fukumura, eds.
doi:10.1017/S1743921323002934

High energy processes in common envelope jets supernovae

Aldana Grichener

Department of Physics, Technion, Haifa, 3200003, Israel
email: aldanag@campus.technion.ac.il

Abstract. We study high energy processes that occur during the merger of a neutron star (NS) or a black hole (BH) with the core of a red supergaint (RSG). The merger powers a luminous event termed common envelope jets supernova (CEJSN), that might account for lightcurves of peculiar transients. In the CEJSN scenario the NS/BH accretes mass from its surroundings through an accretion disk as it spirals-in inside the RSG's envelope and core. The compact object launches part of this mass as narrow jets that interact with their environment by depositing their kinetic energy in the envelope and core gas. These jets can serve as production sites of high energy neutrinos and r-process elements.

Keywords. neutrinos – binaries:general – stars:black holes – stars:jets – stars:massive – transients:supernovae

1. Introduction

The mergers of neutron stars (NSs) and black holes (BHs) with cores of red-supergiants (RSGs) are a subject of extensive research. These mergers might lead to bright transient events that are identified as peculiar supernovae (SNe) explosions. In the *common envelope jets supernova (CEJSN)* scenario for NS/BH-core mergers, a NS/BH is engulfed by an RSG star and spirals-in inside its envelope while accreting and expelling mass in the process. The mass is accreted through an accretion disk that launches two jets that propagate in opposite directions perpendicular to the orbital plane. The compact object eventually reaches the core and launches more energetic jets due to the high accretion rates in the dense core. The deposition of the jets' kinetic energy in the core drives bright transient events.

The most common formation channel of NS/BH core mergers, as Schrøder et al. (2020) and Grichener (2023) find, is as follows. Two massive main sequence (MS) stars in a binary system evolve independently from one another to begin with. At some point, when the initially more massive star is an RSG, mass transfer (either stable or unstable) from the RSG star to its MS companion strips the RSG of its envelope. The naked core keeps evolving and ends its life in a SN explosion, leaving a NS or BH remnant behind. If the system remains bound after the explosion, unstable mass transfer in the opposite direction might lead to a common envelop evolution in which the initially lighter star, which is in its RSG phase at this stage, swallows the compact object, leading to the CEJSN event described above.

Mergers of NS/BH with cores of RSGs might be associated with many peculiar transients. Thöne et al. (2011) suggest that the unusual gamma-ray burst 101225A originated from the merger of an helium star with a NS. Soker & Gilkis (2018) propose a CEJSN scenario to explain the enigmatic iPTF14hls transient. Soker et al. (2019) argue that CEJSNe can explain AT2018cow-like fast blue optical transients. Dong et al. (2021)

propose that merger-driven explosions are responsible for the luminous radio transient VT J121001+495647. Moreover, CEJSN with a NS companion can serve as the production site of heavy r-process elements, as suggested by Papish et al. (2015) and further developed by Grichener & Soker (2019), Grichener et al. (2022) and Grichener & Soker (2022).

In this work we focus on CEJSNe as emission sites of high energy neutrinos. We summarize the main results from Grichener & Soker (2021) in section 2. Then, we compare the properties of different CEJSN scenarios and summarize our results in section 3.

2. High energy neutrinos from common envelope jets supernovae: brief summary of Grichener & Soker (2021)

In Grichener & Soker (2021) we propose and explore a new astrophysical site for the formation and emission of the PeV neutrinos detected by IceCube in 2013 (see Aartsen et al. (2013) for the first observation of PeV-energy neutrinos with IceCube): CEJSN with a BH companion. When the BH is swallowed by an RSG it spirals-in inside its envelope and accretes mass through an accretion disk. The accretion disk launches jets in which high energy neutrinos can be formed. High energy cosmic rays can be scattered from the magnetic filed irregularities inside the jets. The cosmic rays then interact with photons that originate from the jet's head (the contact discontinuity between the reverse-shocked jet and the forward-shocked envelope) and form very energetic and short-lived pions. The charged pions decay to muons and then to electrons, emitting very energetic neutrinos that might be detected by IceCube. The neutral pions decay to gamma-rays. From kinematic considerations, the maximal energy of the high energy neutrinos formed in this process is about 20% of the maximal cosmic rays energy.

We use the stellar evolution code MESA (see Paxton et al. (2019) for the latest version) to evolve a $30M_\odot$ star with solar metallicity $Z_\odot = 0.02$. During its RSG phase, when the star reaches a radius of $1000R_\odot$, we assume that it swallows a $3M_\odot$ BH that begins launching jets while spiralling-in inside the RSG envelope. We use sub-grid physics and mimic the effect of the jets on envelope by injecting the kinetic energy of the jets into the spherically symmetric stellar model. We then extract the density profiles of the extended RSG envelope to perform all the calculations presented in Grichener & Soker (2021).

We compute the maximum energy cosmic rays can reach by acceleration due to the jets' magnetic filed by comparing the typical acceleration timescale to typical cooling timescales. We find that the most relevant cooling process in our scenario is reduction in the protons' energies due to photohadronic interactions. As long as the acceleration timescale is shorter than the cooling timescale, the cosmic rays will keep gaining energy. By equating the acceleration timescale due to magnetic fields in the jets and the cooling timescale due to photohadronic interactions, we find that cosmic rays in our scenario can reach energies of

$$E_{\rm max} \approx 4 \times 10^{16}\ {\rm eV} \left(\frac{{\rm B}'}{10^3\ {\rm G}}\right) \left(\frac{{\rm T}'_\gamma}{0.35\ {\rm keV}}\right)^{-3} \left(\frac{\Gamma}{100}\right), \tag{2.1}$$

where B' is the magnetic filed in the rest frame of the jet, T'_γ is typical temperature of the photons that the cosmic rays interact with as measured in the rest frame of the jet and Γ is the Lorenz factor of the jets. This energy implies that CEJSN with BH companions might account for PeV neutrinos. For more details on the assumptions of our model, the computations we performed and other properties of high energy neutrinos and high energy neutrinos sites that the CEJSN scenario is compatible with please see Grichener & Soker (2021).

Table 1. Summary of the different common envelope jets supernova (CEJSN) scenarios. Based on Table 1 in Soker et al. (2019)

Scenario	Remnant companion	Description	Possible outcomes
CEJSN r-process[PSB15,GS19,GKS22,GS22]	NS	Common envelope evolution of the NS inside the core of the RSG; launching of neutron-rich jets	Heavy r-process nucleosynthesis
Prolonged CEJSN[SG18]	NS/BH	Common envelope evolution of the NS/BH inside the the RSG core; the core remains partly bound after the explosion causing prolonged accretion	iPTF14hls-like peculiar transients
Polar CEJSN[SGG19]	NS/BH	Common envelope evolution of the NS/BH in the RSG core; jets explode the core and eject relativistic polar outflow	AT2018cow-like fast blue optical transients
CEJSN neutrinos [GS21]	BH	Common envelope evolution of the BH inside the RSG envelope; launching of ultra-relativistic jets	Detectable PeV neutrinos
Thermonuclear CEJSN [GS23]	NS	The NS tidally-disrupts the core from the outside; thermonuclear ignition of the accretion disk mater	W49B-like supernova remnants

Refrences: GS19: Grichener & Soker (2019), GKS22: Grichener et al. (2022), GS21: Grichener & Soker (2021), GS22: Grichener & Soker (2022), GS23: Grichener & Soker (2023), PSB15: Papish et al. (2015), SG18: Soker & Gilkis (2018), SGG19: Soker et al. (2019).

3. Summary and discussion

CEJSN is a general name for a class of transients where a NS/BH merges with the core of an RSG star launching jets in the process. In the previous section we elaborated on the CEJSN scenario for high energy neutrino emission and briefly summarized the main findings of Grichener & Soker (2021). In Table 1 we present other CEJSN scenarios we developed over the years along their possible outcomes.

The main conclusion from this proceedings paper, and particularly from Table 1 is that CEJSN might account for several peculiar transients and high energy astrophysical phenomena, which emphasizes the need for more extensive studies in the field.

References

Aartsen, M. G., Abbasi, R., Abdou, Y., et al. 2013, PRL, 111, 021103
Dong, D. Z., Hallinan, G., Nakar, E., et al. 2021, Science, 373, 1125. doi:10.1126/science.abg6037
Grichener, A. & Soker, N. 2019, ApJ, 878, 24. doi:10.3847/1538-4357/ab1d5d
Grichener, A. & Soker, N. 2021, MNRAS, 507, 1651. doi:10.1093/mnras/stab2233
Grichener, A., Kobayashi, C., & Soker, N. 2022, ApJL, 926, L9. doi:10.3847/2041-8213/ac4f68
Grichener, A. & Soker, N. 2022, Research Notes of the American Astronomical Society, 6, 263. doi:10.3847/2515-5172/acaa9f
Grichener, A. & Soker, N. 2023, arXiv:2303.05258. doi:10.48550/arXiv.2303.05258
Grichener, A. 2023, MNRAS, 523, 221. doi:10.1093/mnras/stad1449
Papish, O., Soker, N., & Bukay, I. 2015, MNRAS, 449, 288. doi:10.1093/mnras/stv345
Paxton, B., Smolec, R., Schwab, J., et al. 2019, ApJS, 243, 10,
Schrøder, S. L., MacLeod, M., Loeb, A., et al. 2020, ApJ, 892, 13. doi:10.3847/1538-4357/ab7014
Soker, N. & Gilkis, A. 2018, MNRAS, 475, 1198. doi:10.1093/mnras/stx3287
Soker, N., Grichener, A., & Gilkis, A. 2019, MNRAS, 484, 4972. doi:10.1093/mnras/stz364
Thöne, C. C., de Ugarte Postigo, A., Fryer, C. L., et al. 2011, Nature, 480, 72. doi:10.1038/nature10611

Chapter 5:
Theory & Modelling of outflows

Black Hole Winds at All Scales
Proceedings IAU Symposium No. 378, 2024
G. Bruni, M. Diaz Trigo, S. Laha & K. Fukumura, eds.
doi:10.1017/S1743921323003289

BAL effect in quasars due to source orientation

M. Sniegowska[1,2], **M. H. Naddaf**[1,2], **M. L. Martinez-Aldama**[3,4], **P. Marziani**[5], **S. Panda**[6] **and B. Czerny**[1]

[1]Center for Theoretical Physics, Polish Academy of Sciences,
Lotnikow 32/46, 02-668 Warsaw, Poland
email: msniegowska@camk.edu.pl

[2]Nicolaus Copernicus Astronomical Center, Polish Academy of Sciences,
Bartycka 18, 00-716 Warsaw, Poland
email: naddaf@cft.edu.pl

[3]Instituto de Física y Astronomía, Facultad de Ciencias, Universidad de Valparaíso,
Gran Bretaña 1111, Valparaíso, Chile

[4]Departamento de Astronomía, Universidad de Chile, Casilla 36D, Santiago, Chile

[5]INAF-Astronomical Observatory of Padova, Vicolo dell'Osservatorio, 5,
35122 Padova PD, Italy

[6]Laboratório Nacional de Astrofísica - Rua dos Estados Unidos 154,
Bairro das Nações. CEP 37504-364, Itajubá, MG, Brazil

Abstract. We investigated a scenario where the presence of a broad absorption line (BAL) feature in quasars (QSOs) is contingent upon the line of sight being situated within an outflow cone emanating from the source. We examined the mechanism of dust-driven winds based on the failed radiatively accelerated dusty outflow (FRADO) model proposed by Czerny & Hryniewicz, letting it be responsible for the formation of massive outflow. We calculated the probability of observing the BAL effect from the geometry of outflow which is a function of global parameters of black hole mass ($M_\bullet$), Eddington ratio ($\alpha_{\rm Edd}$), and metallicity (Z). We then compared the results with prevalence of BAL QSOs in a sample of observational data from SDSS. The consistency of our model with the data supports the interpretation of the BAL phenomenon as a result of source orientation, rather than a transitory stage in AGN evolution.

Keywords. Active Galaxies, Broad Absorption Lines, Quasars, Dust, Radiation Pressure

1. Introduction

BAL QSOs represent a unique and enigmatic class of quasars that exhibit prominent blue-shifted absorption features in their spectra (Lynds 1967; Trump et al. 2006) which evidence the presence of strong and massive outflows from the source (Allen et al. 2011; Hamann et al. 2013). They exhibit a continuous and wide absorption trough encompassing velocities reaching several times 10^4 km/s (Risaliti et al. 2005; Gibson et al. 2009). The study of BAL QSOs holds significant importance in understanding the intricate interplay between the accretion disk, black hole, and surrounding environment (Hopkins et al. 2009; Hamann et al. 2019).

BAL QSOs are divided into three types based on the absorption lines present in the spectrum (Hall et al. 2002). High-ionization BAL quasars (HBAL QSOs), constituting approximately 85% of BAL quasars, exhibit absorption features only from highly ionised atoms. Low-ionization BAL quasars (LBAL QSOs) have high-ionization absorption plus displaying absorption troughs in low-ionisation ions, making up approximately 15% of the entire BAL population. FeLBAL QSOs, the rarest subtype among BAL QSOs, exhibit absorption from both high- and low-ionization species, as well as absorption from Iron excited states.

Observationally, a comparison of multiple wavelengths reveals minimal intrinsic distinctions between BALs and non-BALs, except for the tendency of BAL QSOs to exhibit redder ultraviolet continua compared to non-BAL QSOs (Weymann et al. 1991; Lewis et al. 2003). BALs are commonly associated with outflows from the accretion disk of active galactic nuclei (AGNs) (Murray et al. 1995). The origin of BALs in quasars is subject to two theoretical scenarios. The first suggests that BAL QSOs are essentially normal QSOs observed from a line of sight that passes through the outflowing gas (Elvis 2000). The second scenario involves an evolutionary process (Williams et al. 1999) which is no longer preferred (Turnshek 1988). The first scenario relies on the massive outflow of gas due to line-driving mechanism (Elvis 2000; Risaliti et al. 2011). However, there are shreds of evidence in the spectra of BAL sources hinting at the presence of dust (Dunn et al. 2010; Borguet et al. 2013). We, therefore, aimed at testing whether there is any consistency or correlation between the properties of a dust-driven outflow and the probability of observing BAL feature in a given source. For the comprehensive version of this study, see Naddaf et al. (2023).

2. Methodology

We use the enhanced 2.5D version of the failed radiatively accelerated dusty outflow model (FRADO) (Czerny & Hryniewicz 2011; Naddaf & Czerny 2022). FRADO model predicts that radiation pressure acting on dust lifts the clumps of dusty material from the surface of an accretion disk. Depending on the launching location and the global parameters, i.e. black hole mass ($M_\bullet$), Eddington ratio ($\alpha_{\rm Edd}$), and metallicity (Z), the material may then come back to the disk surface or it may escape from the gravitational potential. Also, the launched material may or may not lose the dust content depending on the trajectory (for more details of the 2.5D model, see Naddaf et al. 2021; Naddaf & Czerny 2022). An example of the outflow in the 2.5D FRADO model is shown in figure 1 which in this case includes both highly- and lowly-ionized regions shaded in red (dustless) and blue (dusty), respectively. The geometry of the outflow is a function of $M_\bullet$, $\alpha_{\rm Edd}$, and Z. We calculate the probability of observing BAL feature as

$$P_{\rm BAL} = \frac{\text{fraction of the sky covered by outflow cone}}{\text{unit sphere}} = \cos(\theta_{\rm min}) - \cos(\theta_{\rm max}) \quad (1)$$

As for the observational data, we collected a sample from the quasar catalogue of the Sloan Digital Sky Survey DR7 (Shen et al. 2011). The selected sample consists of sources with a median S/N > 10 per pixel in the rest-frame 2700-2900 Å region in which the black hole masses are estimated from MgII λ2800 . The full sample contains 42,349 objects that 3% of which are detected as BAL QSOs. Black hole masses and Eddington ratios for MgII were obtained from the SDSS catalog. Dividing the parameter space of black hole mass and Eddington ratio, $M_\bullet$-$\alpha_{\rm Edd}$, into bins centered at integer values of log $M_\bullet$ and log$\alpha_{\rm Edd}$with the step size of one in log scale, the median values for the distribution of HBAL and LBAL sources of the sample in each bin are presented in Figure 2. The value from the sample of observational data which can be compared with $P_{\rm BAL}$ from 2.5D FRADO is the *prevalence ratio*, $r_{\rm BAL}$, defined as

$$r_{\rm BAL} = \frac{N_{\rm BAL}}{(N_{\rm BAL} + N_{\rm nonBAL})} \quad (2)$$

where N stands for the corresponding number of sources in each bin.

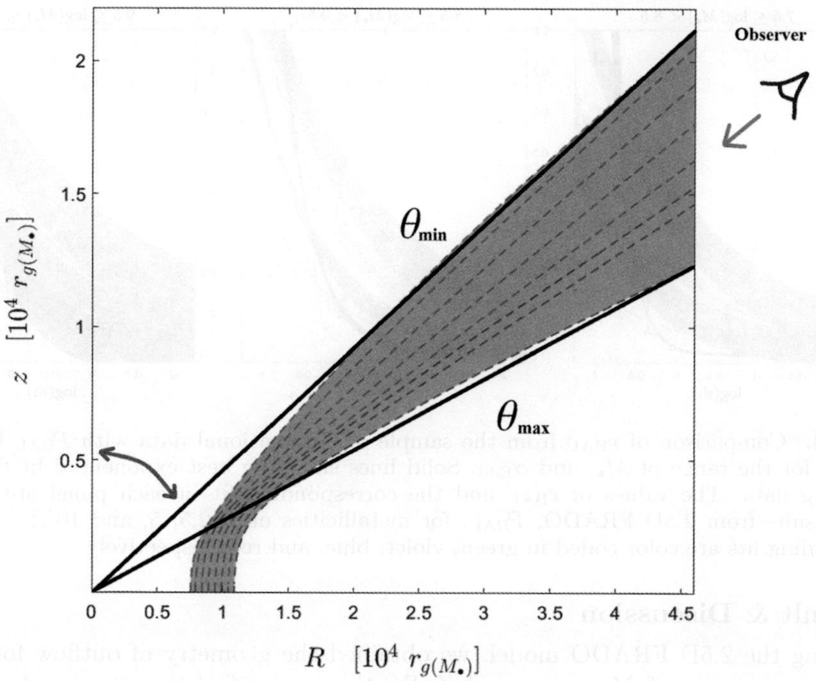

Figure 1. An example of outflow in 2.5D FRADO model for a source at Eddington rate with $M_\bullet$ of $10^8 M_\odot$ and metalicity of $5Z_\odot$. Dotted lines show the path of the trajectory of a clump. The area shaded in blue/red corresponds to the dusty/dustless (lower/higher ionized) actual status of outflowing material. The two black solid lines represent the maximum and minimum angles (to symmetry axis, z) confining the outflow cone. The BAL feature is seen if the line of sight is within the outflow cone.

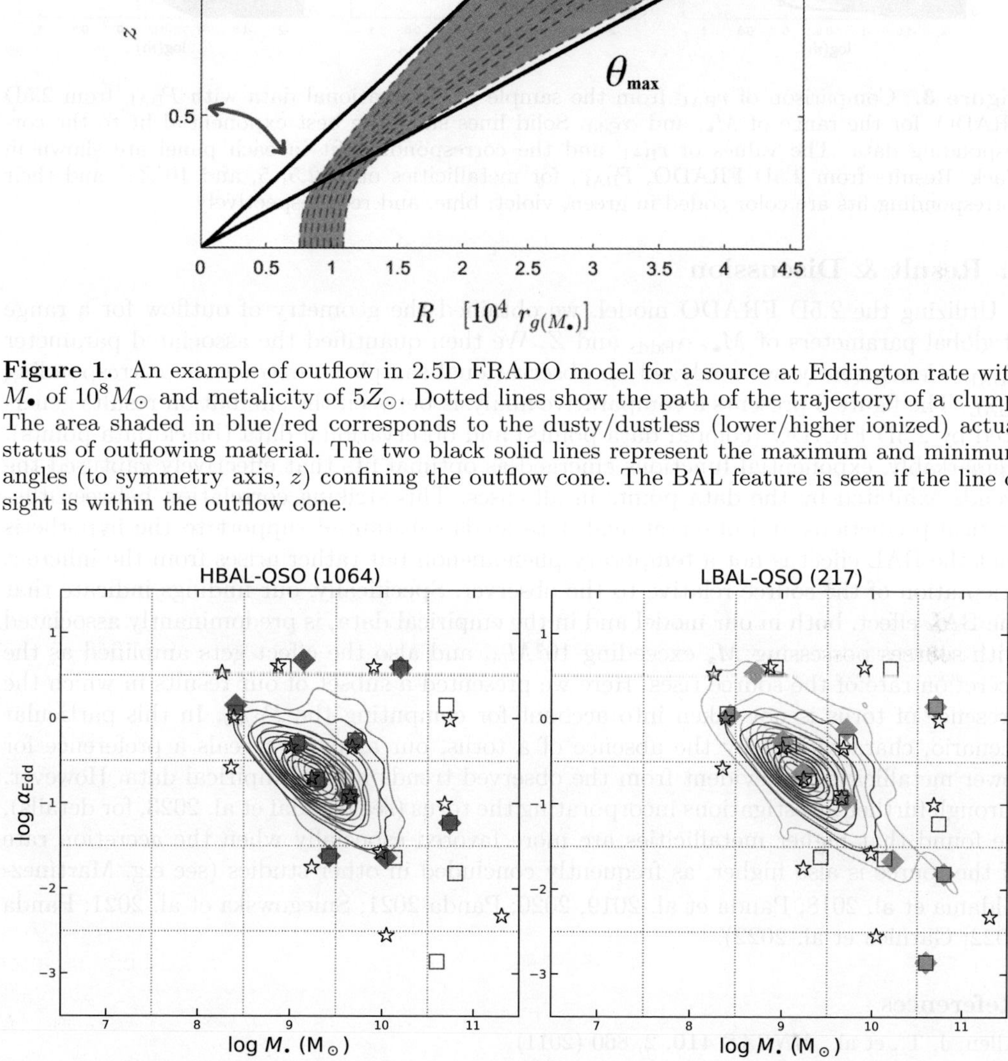

Figure 2. The $M_\bullet$-$\alpha_{\rm Edd}$ space for non-BAL (black $\star$), BAL (blue $\square$), HBAL (orange $\diamond$ in the left panel) and LBAL (green $\diamond$ in the right panel) QSOs. Symbols represent the median values of log $M_\bullet$ and log $\alpha_{\rm Edd}$ in each bin. Contours in black, orange, and green mark the distribution for non-BAL, HBAL, and LBAL QSOs, respectively.

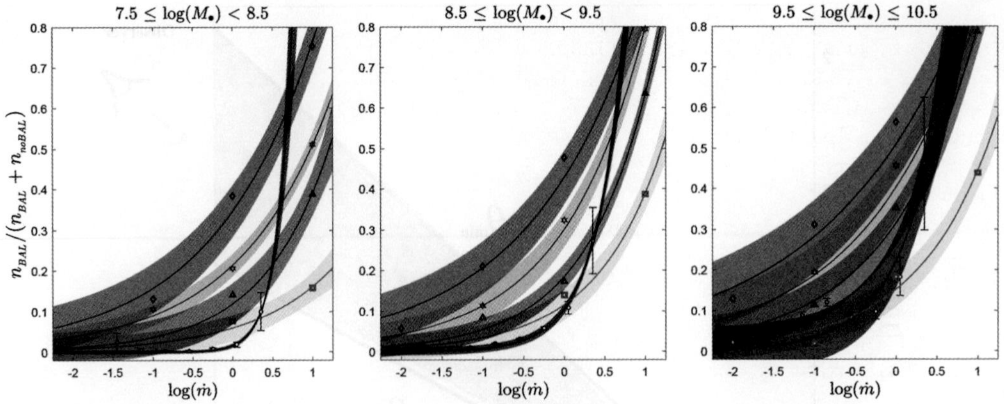

Figure 3. Comparison of $r_{\rm BAL}$ from the sample of observational data with $P_{\rm BAL}$ from 2.5D FRADO, for the range of $M_\bullet$, and $\alpha_{\rm Edd}$. Solid lines show the best exponential fit to the corresponding data. The values of $r_{\rm BAL}$ and the corresponding fit in each panel are shown in black. Results from 2.5D FRADO, $P_{\rm BAL}$, for metallicities of 1, 2.5, 5, and 10 $Z_\odot$, and their corresponding fits are color coded in green, violet, blue, and red, respectively.

3. Result & Discussion

Utilizing the 2.5D FRADO model, we obtained the geometry of outflow for a range of global parameters of $M_\bullet$, $\alpha_{\rm Edd}$, and Z. We then quantified the associated parameter $P_{\rm BAL}$. Concurrently, we analyzed an observational sample to extract the corresponding $r_{\rm BAL}$. The figure 3 presents a comparative analysis between the simulation results generated by 2.5D FRADO (colored data points) and observational data (black data points). Remarkably, exponential functions emerged as optimal fits that effectively captured the trends exhibited by the data points in all cases. This striking correlation between theoretical predictions and observational data lends substantial support to the hypothesis that the BAL effect is not a temporary phenomenon but rather arises from the inherent orientation of the source relative to the observer. Specifically, our findings indicate that the BAL effect, both in our model and in the empirical data, is predominantly associated with sources possessing $M_\bullet$ exceeding $10^8 M_\odot$, and also the effect gets amplified as the accretion rate of the source rises. Here we presented a subset of our results in which the presence of torus is not taken into account for computing the $P_{\rm BAL}$. In this particular scenario, characterized by the absence of a torus, our analysis reveals a preference for lower metallicities, as evident from the observed trends in the empirical data. However, through further investigations incorporating the torus (see Naddaf et al. 2023, for details), we found that higher metallicities are more favored especially when the accretion rate of the source is also higher, as frequently concluded in other studies (see e.g. Martínez-Aldama et al. 2018; Panda et al. 2019, 2020; Panda 2021; Śniegowska et al. 2021; Panda 2022; Garnica et al. 2022).

References

Allen, J. T., et al., *MNRAS* **410**, 2, 860 (2011)
Borguet, B. C. J., et al., *ApJ* **762**, 1, 49 (2013)
Czerny, B., Hryniewicz, K., *A&A* **525**, L8 (2011)
Dunn, J. P., et al., *ApJ* **709**, 2, 611 (2010)
Elvis, M., *ApJ* **545**, 1, 63 (2000)
Garnica, K., et al., *A&A* **667**, A105 (2022)
Gibson, R. R., et al., *ApJ* **692**, 1, 758 (2009)
Hall, P. B., et al., *ApJS* **141**, 2, 267 (2002)

Hamann, F., Herbst, H., Paris, I., Capellupo, D., *MNRAS* **483**, 2, 1808 (2019)
Hamann, F., et al., *MNRAS* **435**, 1, 133 (2013)
Hopkins, P. F., Murray, N., Thompson, T. A., *MNRAS* **398**, 1, 303 (2009)
Lewis, G. F., Chapman, S. C., Kuncic, Z., *ApJL* **596**, 1, L35 (2003)
Lynds, C. R., *ApJ* **147**, 396 (1967)
Martínez-Aldama, M. L., et al., *A&A* **618**, A179 (2018)
Murray, N., Chiang, J., Grossman, S. A., Voit, G. M., *ApJ* **451**, 498 (1995)
Naddaf, M. H., Czerny, B., *A&A* **663**, A77 (2022)
Naddaf, M. H., Czerny, B., Szczerba, R., *ApJ* **920**, 1, 30 (2021)
Naddaf, M. H., et al., *A&A* (2023)
Panda, S., *A&A* **650**, A154 (2021)
Panda, S., *Frontiers in Astronomy and Space Sciences* **9**, 850409 (2022)
Panda, S., Marziani, P., Czerny, B., *ApJ* **882**, 2, 79 (2019)
Panda, S., Marziani, P., Czerny, B., *Contributions of the Astronomical Observatory Skalnate Pleso* **50**, 1, 293 (2020)
Risaliti, G., et al., *ApJL* **630**, 2, L129 (2005)
Risaliti, G., et al., *MNRAS* **410**, 2, 1027 (2011)
Shen, Y., et al., *ApJS* **194**, 2, 45 (2011)
Śniegowska, M., et al., *ApJ* **910**, 2, 115 (2021)
Trump, J. R., et al., *ApJS* **165**, 1, 1 (2006)
Turnshek, D. A., QSO Absorption Lines, Cambridge Uni. Press (1988)
Weymann, R. J., Morris, S. L., Foltz, C. B., Hewett, P. C., *ApJ* **373**, 23 (1991)
Williams, R. J. R., Baker, A. C., Perry, J. J., *MNRAS* **310**, 4, 913 (1999)

Black Hole Winds at All Scales
Proceedings IAU Symposium No. 378, 2024
G. Bruni, M. Diaz Trigo, S. Laha & K. Fukumura, eds.
doi:10.1017/S1743921323003150

Photon scattering in a relativistic outflow having velocity shear: a novel mechanism of generation for high energy power-law spectra

Mukesh Kumar Vyas and Asaf Pe'er

Bar Ilan University, Ramat Gan, Israel. 5290002
email: mukeshkvys@gmail.com

Abstract. We show that an extragalactic jet with a velocity shear gives rise to Fermi like acceleration process for photons scattering withing the shear layers of the jet. Such photons then gain energy to produce a high energy power law. These power law spectra at high energies are frequently observed in several extragalactic objects such as Gamma Ray Bursts (GRBs). We implement the model on GRBs to show that the obtained range of the photon indices are well within their observed values. The analytic results are confirmed with numerical simulations following Monte Carlo approach.

Keywords. Relativistic jets, High energy astrophysics, Gamma-ray bursts, Theoretical models

1. Introduction

Many astrophysical objects show a power law component in their high energy spectrum like gamma-ray bursts [GRBs] (Band et al. 1993; Kaneko et al. 2006; Bošnjak et al. 2014; Pe'er 2015; Preece et al. 2000; Barraud et al. 2003), Active galactic nuclei [AGNs] (Nandra & Pounds 1994; Reeves & Turner 2000; Page et al. 2005) etc. They harbour a relativistic jet propagating with a Lorentz factor Γ. The existence of high energy power law in them is associated with processes like Synchrotron Band et al. (1993); Kaneko et al. (2006) or Compton scattering (Blumenthal & Gould 1970; Rybicki & Lightman 1986; Zdziarski & Krolik 1993; Vyas et al. 2021a,b).

Here we present an alternate aspect of generation of power law spectra in a relativistic flow. Considering that a relativistic jet with a differential Lorentz factor within its layers is a generic picture, we show that a photon scattering between these layers can gain or lose energy between two scattering events with electrons. However, when the photon goes through several scatterings, on average, it leads to an energy gain. A population of photons scattering within and escaping to an observer is shown to distribute like a power law spectrum at high energies. We argue that this mechanism is a viable candidate to generate high energy power-law spectra. Further, from the observations, this model hands us a tool to infer the jet structures and respective Lorentz factor profile.

2. Generation of power law spectra in relativistic jets

Consider a population of photons injected deep inside a jet and left to scatter between shear layers before escaping. The jet assumes an angle dependent Lorentz factor profile due to shear that developed as the jet crosses the progenitor star. It is given as,

$$\gamma(\theta) = \gamma_{\min} + \frac{\gamma_0}{\sqrt{\left(\frac{\theta}{\theta_{\rm j}}\right)^{2p} + 1}}. \tag{1}$$

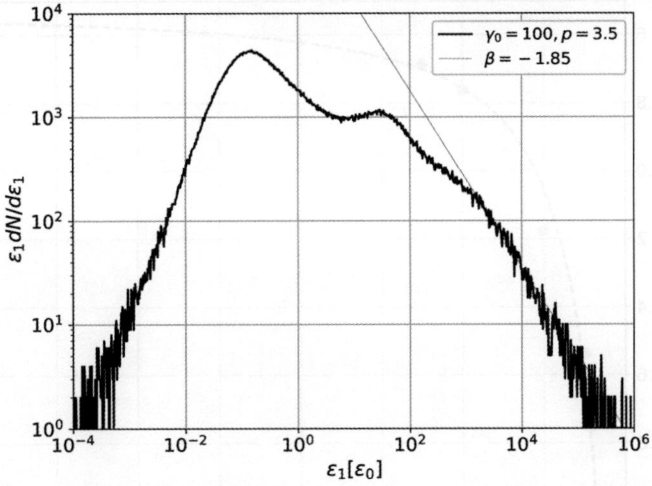

Figure 1. Numerical results of the spectrum produced with parameters $\gamma_0 = 100, p = 3.5, \theta_j = 0.01$.

Here γ_0 is the maximum Lorentz factor of the jet along its axis and it decays like $\gamma \propto \theta^{-p}$ beyond $\theta > \theta_{\rm j}$. At large angles, $\theta >> \theta_{\rm j}$, Lorentz factor settles at $\gamma_{\rm min}$. To estimate the spectrum, we follow the following procedure to obtain energy evolution of the photons in the jet. After k scattering N photons are left within the region out of N_0. On average, a photon has energy $\varepsilon_k = \varepsilon_i \bar{g}^k$ if it was injected with energy ε_i. Here $\bar{g}$ is defined as the average energy gain per scattering. This implies,

$$\frac{N}{N_0} = \left[\frac{\varepsilon_k}{\varepsilon_i}\right]^{\beta} \tag{2}$$

with $\beta' = \frac{\ln P}{\ln \bar{g}}$ and the photon index β is obtained as

$$\beta = \beta' - 1 = \frac{\ln \bar{P}}{\ln \bar{g}} - 1. \tag{3}$$

Here $\bar{P}$ is the average probability of scattering without escape. The details of obtaining scattering probabilities, as well as the average energy gain for a given Lorentz factor profile are described in Vyas & Pe'er (2023).

Alongside the analytic approach described above, we also perform numerical simulations to validate the theory. We inject photons in our simulation code that tracks and counts photons that are emitted towards certain observer with escape given escape energy. The photon population thus produce the spectrum for the escaped photons. For the details of Monte Carlo simulation code, see Vyas et al. (2021a,b); Pe'er (2008); Lundman et al. (2013).

3. Results

The photons are injected in a jet with parameters $\gamma_0 = 100, \theta_{\rm j} = 0.01$ rad, $\gamma_{\rm min} = 1.001$ and profile index $p = 3.5$ and the produced spectrum using Monte Carlo simulations is shown in Figure 1. This spectrum is seen by an on axis observer situated at a polar angle $\theta_{\rm o} = 0$. The photons are injected with seed energy $\varepsilon_0 = 10^{-10}$ erg and scattered energies ε_1 are distributed in the spectra indicating a high energy power law beyond energies $\varepsilon_1/\varepsilon_0 > 10^3$. The simulated photon index $\beta = -1.85$ is confirmed with the analytic model presented in the previous section. In Figure 2, we keep all parameters to be the same and change the profile index p from 2.5 to 10 and plot the photon indices β

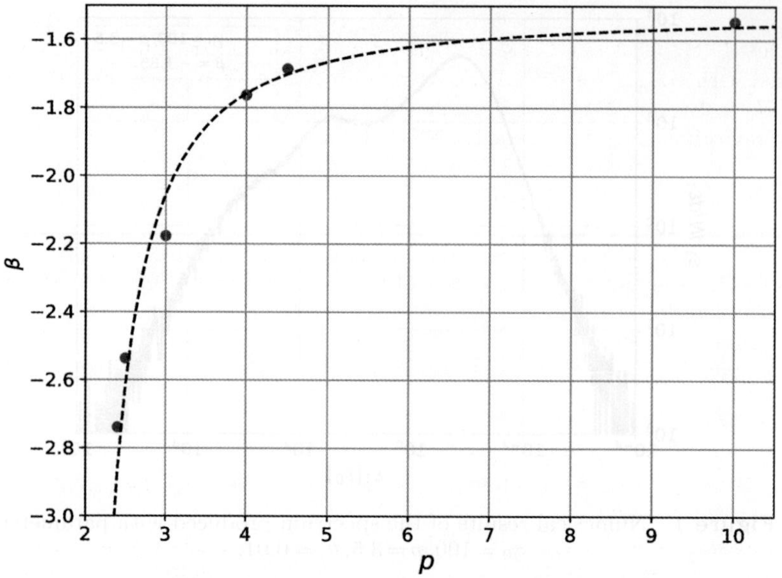

Figure 2. Distribution of photon index β with p in range $2-10$. Other parameters are same as Figure 1.

with p. The analytic curve is shown by dashed lines while the respective simulated results are overplotted with red dots. The spectra gets harder as p increases and saturates at $\beta = -1.55$ as $p \to \infty$. At small values of p, the power law spectra vanishes (for $p < 2.5$). It happens due to adiabatic energy losses because the shear in jet is not sufficient to counter radial expansion and the photons do not gain energy. The agreement between the simulations and theoretical results confirm the theoretical model developed.

4. Conclusions

We have simulated photons that undergo scattering within the shear layers of a relativistic jet and shown that the net energy gain of these photons is capable to produce a high energy power-law spectrum. This mechanism is similar to Fermi acceleration for charged particles in the presence of magnetic fields (Blandford & Eichler 1987). We discuss the same mechanism for the case of photons in shearing media with relativistic electrons. The Fermi acceleration for electrons is widely explored while we work out the photon acceleration and show that the mechanism has wide applications. The obtained range of the photon indices here is well within the range obtained in GRB observations at high energies between -5 to -1.5 (Preece et al. 1998; Kaneko et al. 2006).

In further works, we will work out the implications of this mechanism in other astrophysical objects such as jets in active galactic nuclei (AGNs) and accretion discs around black holes.

References

Band, D., Matteson, J., Ford, L., Schaefer, B., Palmer, D., Teegarden, B., Cline, T., Briggs, M., Paciesas, W., Pendleton, G., Fishman, G., Kouveliotou, C., Meegan, C., Wilson, R., & Lestrade, P. 1993, BATSE Observations of Gamma-Ray Burst Spectra. I. Spectral Diversity, The Astrophysical Journal, 413, 281.

Barraud, C., Olive, J. F., Lestrade, J. P., Atteia, J. L., Hurley, K., Ricker, G., Lamb, D. Q., Kawai, N., Boer, M., Dezalay, J. P., Pizzichini, G., Vanderspek, R., Crew, G., Doty, J., Monnelly, G., Villasenor, J., Butler, N., Levine, A., Yoshida, A., Shirasaki, Y., Sakamoto,

T., Tamagawa, T., Torii, K., Matsuoka, M., Fenimore, E. E., Galassi, M., Tavenner, T., Donaghy, T. Q., Graziani, C., & Jernigan, J. G. 2003, Spectral analysis of 35 GRBs/XRFs observed with HETE-2/FREGATE, Astronomy and Astrophysics, 400, 1021–1030.

Blandford, R. & Eichler, D. 1987, Particle acceleration at astrophysical shocks: A theory of cosmic ray origin, Physics Reports, 154(1), 1–75.

Blumenthal, G. R. & Gould, R. J. 1970, Bremsstrahlung, Synchrotron Radiation, and Compton Scattering of High-Energy Electrons Traversing Dilute Gases, Reviews of Modern Physics, 42(2), 237–271.

Bošnjak, Ž., Götz, D., Bouchet, L., Schanne, S., & Cordier, B. 2014, The spectral catalogue of INTEGRAL gamma-ray bursts. results of the joint IBIS/SPI spectral analysis, Astronomy and Astrophysics, 561, A25.

Kaneko, Y., Preece, R. D., Briggs, M. S., Paciesas, W. S., Meegan, C. A., & Band, David L., a. 2006, The Complete Spectral Catalog of Bright BATSE Gamma-Ray Bursts, 166(1), 298–340.

Lundman, C., Pe'er, A., & Ryde, F. 2013, A theory of photospheric emission from relativistic, collimated outflows, MNRAS, 428(3), 2430–2442.

Nandra, K. & Pounds, K. A. 1994, GINGA observations of the X-ray spectra of Seyfert galaxies, MNRAS, 268, 405–429.

Page, K. L., Reeves, J. N., O'Brien, P. T., & Turner, M. J. L. 2005, XMM-Newton spectroscopy of high-redshift quasars, MNRAS, 364(1), 195–207.

Pe'er, A. 2008, Temporal Evolution of Thermal Emission from Relativistically Expanding Plasma, Astrophysical Journal, 682(1), 463–473.

Pe'er, A. 2015, Physics of Gamma-Ray Bursts Prompt Emission, Advances in Astronomy, 2015, 907321.

Preece, R. D., Briggs, M. S., Mallozzi, R. S., Pendleton, G. N., Paciesas, W. S., & Band, D. L. 2000, The BATSE Gamma-Ray Burst Spectral Catalog. I. High Time Resolution Spectroscopy of Bright Bursts Using High Energy Resolution Data, Astrophysical Journals, 126(1), 19–36.

Preece, R. D., Pendleton, G. N., Briggs, M. S., Mallozzi, R. S., Paciesas, W. S., Band, D. L., Matteson, J. L., & Meegan, C. A. 1998, BATSE Observations of Gamma-Ray Burst Spectra. IV. Time-resolved High-Energy Spectroscopy, Astrophysical Journal, 496(2), 849–862.

Reeves, J. N. & Turner, M. J. L. 2000, X-ray spectra of a large sample of quasars with ASCA, MNRAS, 316(2), 234–248.

Rybicki, G. B. & Lightman, A. P. 1986,. *Radiative Processes in Astrophysics.*

Vyas, M. K. & Pe'er, A. 2023, Photons' Scattering in Relativistic Plasma with Velocity Shear: Generation of High Energy Power-law Spectra, Astrophysical Journal, Letters, 943(1), L3.

Vyas, M. K., Pe'er, A., & Eichler, D. 2021,a A Backscattering-dominated Prompt Emission Model for the Prompt Phase of Gamma-Ray Bursts, Astrophysical Journal, 908a(1), 9.

Vyas, M. K., Pe'er, A., & Eichler, D. 2021,b Predicting Spectral Parameters in the Backscattering-dominated Model for the Prompt Phase of GRBs, Astrophysical Journal, Letters, 918b(1), L12.

Zdziarski, A. A. & Krolik, J. H. 1993, Compton Scattering and the Gamma-Ray Power-Law Spectrum in Markarian 421, Astrophysical Journal, Letters, 409, L33.

Black Hole Winds at All Scales
Proceedings IAU Symposium No. 378, 2024
G. Bruni, M. Diaz Trigo, S. Laha & K. Fukumura, eds.
doi:10.1017/S1743921323003241

Outflows from black hole accretion flows with various accretion rates

De-Fu Bu and Feng Yuan

Shanghai Astronomical Observatory, Chinese Academy of Sciences,
80 Nandan Road, Shanghai 200030, China
email: dfbu@shao.ac.cn, fyuan@shao.ac.cn

Abstract. Both observations and theoretical studies have convincingly shown that outflows (i.e., wind and jet) are common phenomena from black hole accretion systems with various accretion rates, although the physical driving mechanisms are not exactly same for different accretion modes. Outflows are not only important in the dynamics of black hole accretion, but also play an important role in AGN feedback; therefore it is crucial to investigate their main physical properties including mass flux and velocity. In this paper we summarize recent studies in investigating the properties and driving mechanisms from black hole accretion flows with various accretion rates.

Keywords. accretion, accretion discs, black hole physics, hydrodynamics

1. Introduction

Depending on the value of the accretion rate, black hole accretion flow can be divided into two modes, i.e., the hot accretion flow and cold accretion disk (Yuan & Narayan 2014). The cold flow can be further divided into the standard thin disk and super-Eddington accretion flow, bounded by the Eddington accretion rate.

For hot accretion flow, due to the high specific energy of gas, outflow is speculated to be present (Narayan & Yi 1994). The pioneer global numerical simulation of hot accretion flow show that the mass accretion rate decreases inwards (Stone et al. 1999). Two models have been proposed to interpret this result. In the convection-dominated accretion flow (CDAF; Narayan et al. 2000), it is believed that the convective motion of the accretion flow results in the inward decrease of accretion rate. In the alternative adiabatic inflow-outflow model, it is believed that strong wind is present and takes away mass (ADIOS; Blandford & Begelman 1999). To solve this puzzle, Yuan et al. (2012; see also Narayan et al. 2012) have performed numerical simulations of accretion flows and analyzed the simulation data. They systematically compare the properties of "inflow" and "outflow" and find that the properties of "outflow" is systematically different from those of "inflow". Moreover, it is found that the magnetized hot accretion flow is convectively stable. These results make Yuan et al. (2012) conclude that strong wind must be present. For the cold thin disk, both the line force and magnetic field can drive strong wind (e.g., Proga et al. 2000). Super-Eddington accretion flow is believed to have strong outflows as well (e.g., Sadowski et al. 2016).

Wind plays an important role in the physics of black hole accretion because it can transfer angular momentum and change the density profile of accretion flow and subsequently its radiation. Moreover, recent studies find that wind likely plays a crucial role in active galactic nuclei feedback (e.g., Yuan et al. 2018). In the following, we review the current studies of wind properties and driving mechanisms for both hot and cold accretion modes.

2. Outflows from black hole accretion flow

2.1. *Outflows from hot accretion flow*

We have studied the properties of outflows from hot accretion flows by performing GRMHD simulations of both SANE and MAD around black holes with different spins, i.e., SANE00 ($a=0$), SANE98 ($a=0.98$), MAD00 ($a=0$) and MAD98 ($a=0.98$) (Yuan et al. 2015; Yang et al. 2021). We find that winds are always present regardless of the black hole spin and magnetic field strength. The mass flux of wind increases with radius,

$$\dot{M}_{\rm w} = \left(\frac{r}{R_0}\right)^s \dot{M}_{\rm BH} \quad (2.1)$$

where $\dot{M}_{\rm BH}$ is the accretion rate at the black hole horizon. For models SANE00, SANE98, MAD00, and MAD98, the values of (R_0, s)= ($20r_s$, 1), ($15r_s$, 1.16), ($55r_s$, 1.54), ($30r_s$, 1.26), respectively, with r_s being the Schwarzschild radius. Therefore, given a same black hole accretion rate and beyond a certain radius, the mass flux of wind is stronger when the magnetic field is stronger. This is because the magnetic pressure gradient force is found to be important in driving wind. We would like to mention that the mass flux is obtained by using the trajectory approach, i.e, we follow the motion of the wind particles. The usually adopted "time-average" approach in literature is found to underestimate the mass flux by a factor of 2-6 (Yuan et al. 2015; Yang et al. 2021, 2023).

The mass flux-weighted poloidal velocity of wind $v_{\rm p}(r)$ as a function of radius r is found to be proportional to the local Keplerian speed $v_{\rm K}$, i.e., $v_{\rm p}=\alpha v_{\rm K}$, with α being in the range of $0.2-0.66$. The decrease of the velocity with increasing radius is because wind produced at larger radii has smaller velocities (Yuan et al. 2015). The Bernoulli parameter of wind is not a constant, but keep increasing outward, which is because the accretion flow is not in strictly steady state and inviscid, but turbulent (Yuan et al. 2015).

It is found that wind can only be produced within the outer boundary of accretion flow, i.e., the Bondi radius (Bu 2016). The large-scale dynamics of wind beyond the outer boundary of accretion flow is studied in Cui & Yuan (2020) and found that wind can escape the gravity of the black hole and reach large scales.

Yang et al. (2021) also studied the properties of jet powered by the Blandford-Znajek mechanism and made comparison with wind. They find that the momentum flux of wind is comparable to that of jet, but the power of jet is roughly an order of magnitude higher than wind. These results, combined with the fact that the opening angle of wind is much larger than that of jet, imply the importance of wind in AGN feedback. The driven mechanism of wind and jet is also studied in Yuan et al. (2015) and Yang et al. (2021). Wind is found to be driven by a combination of gas pressure gradient force, magnetic pressure gradient force and centrifugal force; while jet is mainly accelerated by the magnetic pressure gradient force.

There are growing observational evidences for wind from hot accretion flows (see reviews in Yang et al. 2021). Observations of the low-luminosity active galactic nuclei M81 find blue-shifted and red-shifted iron emissions lines. By running MHD numerical simulations and comparing to the observational data, Shi et al. (2021) find that the emission lines come from wind launched from the hot accretion flow in M81, providing the first direct evidence of wind. As an example of wind-ISM interaction, the *Fermi* bubbles detected in the Galactic center may be inflated by wind from the hot accretion flow in Sgr A* (Mou et al. 2014).

2.2. *Wind from a cold thin disk*

In the MHD disk wind model, in order to simplify the equations, isothermal or adiabatic equation of states is usually adopted in literature. Wang et al. (2022) recently study the wind production from a thin disk by performing two-dimensional MHD simulations based on realistic energy equations. They find that the wind properties strongly depend on the treatment of thermodynamics. The isothermal and adiabatic models underestimate the wind power and overestimate the wind temperature. Additionally, the isothermal and adiabatic assumptions erase the local structure of disc winds.

Wang et al. 2022 also compare their simulation results with the observations of UFOs. The observed UFOs have velocity higher than 10^4 km/s; the ionization parameter is in the range $2.5 < \log \xi < 5.5$; the column density is in the range $22 < \log N_H < 24$. Not all winds found by Wang et al. (2022) meet the conditions of the observed UFOs. They pick up wind that satisfy the conditions of UFOs and find that in the model considering realistic radiative cooling and heating, the wind that satisfying the condition of UFOs covers the region between $\theta = 4.7$ and $45.6°$. The corresponding solid angle $\Omega \sim 3.85$; the fraction of Ω to a whole sphere, $\Omega/4\pi \sim 30.6$ percent. It is close to the lower limit of the fraction of UFO in AGNs.

Wang et al. (2022) also calculate the mass flux, momentum flux, and kinetic power of UFOs found in their simulations and compare them to observations of UFOs. They find that inside $600 r_s$, the mass flux, momentum flux and kinetic power of UFOs found in the simulations are well consistent with those of the observed UFOs. However, a cutoff at 600 r_s is found. Beyond $600 r_s$, there is no wind in the simulations satisfying the condition of the observed UFOs. Combining the ionization parameter and the temperature of our simulations, they find that the reason of the missing UFOs at large radii is the ionization state is too low. On the other hand, it has been reported that the UFOs can exist at $10^2 - 10^4 r_s$ (Gofford et al. 2015). Thus, some physics might be missing in MHD disc winds models. A plausible conjecture for the missing physics would be line force. As the winds are sparse and cool, line force might operate on the winds to change their properties. It is thus important to study wind by combining magnetic field and line force.

2.3. *Outflows from super-Eddington accretion flows*

Using the virtual particle trajectory approach proposed in Yuan et al. (2015), Yang et al. (2023) have investigated the properties of wind and jet launched from a magnetically arrested accretion flow with $\dot{M}_{\rm BH} = 12 \dot{M}_{\rm Edd}$ around a black hole with $M_{\rm BH} = 10^6 M_\odot$ and $a = 0.8$. Consistent with previous studies, they find that strong wind can be launched from the super-Eddington accretion flow. Due to the presence of outflow, the mass accretion rate decreases with decreasing radius $\dot{M} \propto r^{0.39}$. The mass flux of wind as a function of accretion rate and radius can be described by,

$$\dot{M}_{\rm w} = \left(\frac{r}{45 r_s}\right)^{0.83} \dot{M}_{\rm BH} \tag{2.2}$$

Different from the case of hot accretion flows, the velocity of wind is found to be a constant of radius, which is ∼15 percent of the speed of light. Such a difference is because of the additional acceleration force in the case of super-Eddington accretion flow, i.e., the radiation force. They have also compared the momentum flux and kinetic power of wind with jet and found that the momentum flux of wind is 1.2 times that of jet while the kinetic power of jet is 2.5 times of that of wind.

The driving mechanism of wind and jet is also studied in Yang et al. (2023). In the case of jet, the force is fully dominated by the Lorentz force. In the case of wind, the results are a bit complicated, depending on locations. In the wind region close to the surface of the accretion flow, the optical depth is large and the radiation and gas are strongly coupled, so the radiation force and the Lorentz force are comparable and both of them play important roles in the acceleration of wind. Well above the main body of the accretion flow, the optical depth becomes smaller. In this location, compared to the Lorentz force, the radiation force becomes less important in accelerating the wind and is the secondarily important force.

3. Summary

The driving mechanism and main properties of wind and jet are studied by performing (radiative) MHD numerical simulations of accretion flows with various accretion rates around black holes with various spin values. Such studies are performed by the trajectory approach, which can give more exact result compared to the widely adopted time average approach.

For hot accretion flow, the mass flux and velocity as a function of black hole accretion rate and black hole spin for both SANE and MAD are presented and compared to jet. The wind is accelerated by the combination of gas pressure gradient force, magnetic pressure gradient force and centrifugal force; while jet is mainly accelerated by the magnetic pressure gradient force.

In the case of wind launched from a thin disk, we only focus on magnetically driven wind. We find that it is important to adopt a realistic energy equation. In addition, the MHD wind model alone can not interpret the observed ultra-fast outflows, suggesting that we need to combine other mechanisms such as line force.

Strong wind is found in the case of super-Eddington accretion flows and the main properties of wind are presented and compared to jets. The speed of wind is found to be roughly a constant and independent of radius. Lorentz force is found to play an important role in accelerating wind and jet. Close to the surface of the accretion flow, radiation force is found to play a comparable role with Lorentz force in driving wind.

References

Blandford, R., & Begelman, M. 1999, *MNRAS*, 303, L1
Bu, D., Yuan, F., Gan, Z., & Yang, X. 2016, *ApJ*, 823, 90
Cui, C., & Yuan, F. 2020, *ApJ*, 890, 81
Gofford, J. et al. 2015, *MNRAS*, 451, 4169
Mou, G., Yuan, F., Bu, D., Sun, M., & Su, M. 2014, *ApJ*, 790, 109
Narayan, R., & Yi, I. 1994, *ApJ*, 428, L13
Narayan, R., Igumenshchev, I. V., & Abramowicz, M. A. 2000, *ApJ*, 539, 798
Narayan, R., Sadowski, A., Penna, R. F., & Kulkarni, A. K. 2012, *MNRAS*, 426, 3241
Proga, D., Stone, J. M., & Kallman, T. R. 2000, *ApJ*, 543, 686
Sadowski, A., Lasota, J., Abramowicz, M. A. & Narayan, R. 2016, *MNRAS*, 456, 3915
Shi, F., Li, Z., Yuan, F., & Zhu, B. 2021, *NatAs*, 5, 928S
Stone, J. M., Pringle, J. E., & Begelman, M. 1999, *MNRAS*, 310, 1002
Wang, W., Bu, D., & Yuan, F. 2022, *MNRAS*, 513, 5818
Yang, H., Yuan, F., Yuan, Y., & White, C. J. 2021, *ApJ*, 914, 131
Yang, H., Yuan, F., Kwan, T., & Dai, L. 2023, *MNRAS*, arXiv:2211.10710
Yuan, F., Bu, D., & Wu, M. 2012, *ApJ*, 761, 130
Yuan, F., & Narayan, R. 2014, *ARA&A*, 52, 529
Yuan, F., Gan, Z., Narayan, R., Sadowski A., Bu, D., & Bai, X. 2015, *ApJ*, 804, 101
Yuan, F., Yoon D., Li, Y., Gan, Z., Ho, L. C., Guo, F. 2018, *ApJ*, 857, 121

Black Hole Winds at All Scales
Proceedings IAU Symposium No. 378, 2024
G. Bruni, M. Diaz Trigo, S. Laha & K. Fukumura, eds.
doi:10.1017/S1743921323002867

Updated Calculations of the Spectral-Line Radiation Force & Mass-Loss Rates for AGN Outflows

Aylecia S. Lattimer and Steven Cranmer

Department of Astrophysical & Planetary Sciences, Laboratory for Atmospheric and Space Physics, University of Colorado, Boulder, CO 80309, USA
email: aylecia.lattimer@colorado.edu
email: Steven.Cranmer@lasp.colorado.edu

Abstract. Photon-driven flows have been studied for almost a century, and a quantitative description of the radiative forces on atoms and ions is important for understanding a wide variety of systems, including active galactic nuclei (AGN). The colloquially-termed "radiation pressure" of line-driven winds plays an important role in driving outflows in these environments. Quantifying the associated forces is crucial to understanding how these flows enable interactive mechanisms within these environments, such as AGN feedback. Here we provide new calculations of the dimensionless line strength parameter due to radiation driving. For representative AGN, we calculate the photoionization balance at each step along the line of sight (LOS) to the proposed wind-launching region above the accretion disk. We then use a recently compiled database of approximately 5.6 million spectral lines to compute the strength of the line-driving force on the gas and the mass-loss rates resulting from these outflows. We also introduce a "shielding factor that increases the magnitude of the accretion disk column density prior to the launch radius. This shielding factor simulates a proposed inner failed wind region that is thought to shield the outflowing gas from becoming over-ionized by the central source. We also revisit and formalize the role of the commonly-used ionization parameter in setting the properties of the accelerating gas.

Keywords. AGN outflows, radiative processes, atomic physics

1. Introduction

Active Galactic Nuclei (AGN) are believed to be powered by accretion onto a supermassive black hole (SMBH), and quantifying and understanding the associated mass and energy flows are crucial to understanding the observed properties of quasars, blazars, Seyfert, and radio galaxies, as well as how these objects evolve and interact with their surroundings. These outflows are suspected of influencing star formation rates far beyond the small-scale size of the AGN itself. This long-range"feedback" is likely responsible for producing several known correlations between the SMBH mass and a range of large-scale properties of the host galaxy and intergalactic medium (Magorrian et al. 1998; Silk & Rees 1998; Cavaliere et al. 2002; Ostriker et al. 2010; Fabian 2012; King & Pounds 2015; Hopkins et al. 2016; Harrison et al. 2018).

There are multiple possible mechanisms to accelerate AGN winds. It is likely that some combination of effects may be responsible for accelerating various features. However, this work focuses on radiative driving as a possible mechanism for launching and sustaining AGN outflows (Giustini & Proga 2019). This is often referred to as "radiation pressure," where the force of radiation from the spectral lines acts on the material of the outflow.

The extremely large number of spectral lines in any given ion has a dominant effect on this pressure on the flow material (Castor 1974; Castor et al. 1975). The absorption and re-emission of photons in a spectral line results in a radial transfer of momentum, driving a wind outward from the source, in this case an AGN. This type of driving has been suggested to occur in the broad absorption line (BAL) region of AGN, where the blueshift of the spectral lines suggest that outflow velocities can reach up to $0.2c$. Line-driven disk winds are a promising hydrodynamical scenario for these outflows, due in part to the large number of strong resonance lines present in the spectra of many BAL quasars (BALQSOs; Murray et al. 1995; Proga et al. 2000; Proga 2007; Risaliti & Elvis 2010; Higginbottom et al. 2014; Nomura et al. 2016; Zhu et al. 2022).

However, in order for line-driving to launch and maintain an outflow from the accretion disk, the material must be relatively lowly-ionized. As more of the wind material becomes highly ionized, such as that in close proximity to the black hole, there are fewer spectral lines present for the radiation from the source to act upon (Higginbottom et al. 2014). A proposed solution to this issue is a so-called "failed wind" near the interior of the accretion disk, which acts to shield the outer material from becoming over-ionized (see, for example, Proga et al. 2000, Proga & Kallman 2004).

In this work, we provide new calculations of the dimensionless line strength parameter due to radiation driving, as well as explore the importance of the previously mentioned failed wind in launching and maintaining a line-driven disk wind. We also revisit and formalize the role of the commonly-used ionization parameter in setting the properties of the accelerating gas.

2. Overview

Since a complete tabulation of the line force requires the inclusion of as many lines as possible, the atomic line data used in this work was retrieved from a variety of sources. Multiple atomic databases were used to fill in any gaps in the available data wherever possible. Databases used include the National Institute of Standards and Technology (NIST; Kramida et al. 2018), version 10.0 of the CHIANTI database (Dere et al. 1997; Del Zanna et al. 2021), the database of lines used by the radiative transfer code CMFGEN (Hillier 1990; Hillier & Miller 1998; Hillier & Lanz 2001), the Opacity Project's TOPbase (Cunto & Mendoza 1992; Cunto et al. 1993), and the XSTAR database (Mendoza et al. 2021). For the AGN central sources considered here the more highly ionized states of each element become increasingly important. Therefore, where available, all ionization states for all elements up to Zn ($Z = 30$) were included. The CHIANTI and CMFGEN databases additionally include both theoretical and observed spectral line data. For the sake of completeness, the compiled line list is comprised of both theoretical (where applicable) and observed transitions. The final line list contained data for 5,671,602 spectral lines.

The core aim of this work is to determine the usefulness of the ionization parameter ξ in determining the ionization balance of the wind material. Since our goal is to model line forces for a wide range of possible AGN environments and outflow features, we vary the basic properties of the AGN and the absorbed line of sight in order to reveal the full range of possible regimes. The initial input parameters of each model include mass of the black hole, Eddington ratio, wind mass flux and speed, wind launch radius, viewing angle, and shielding factor, which enhances the hydrogen number density prior to the launch radius. To determine the efficacy of the ionization parameter ξ, we calculate the complete ionization balance at every step along the modeled line of sight. The model geometry used is shown in Figure 1. The initial spectral energy distribution (SED) is given by the python module QSOSED (`https://github.com/arnauqb/qsosed`). We calculate the initial thermal equilibrium and ionization balance from the SED at the first radial step, defined as the radius of the corona, R_{cor}.

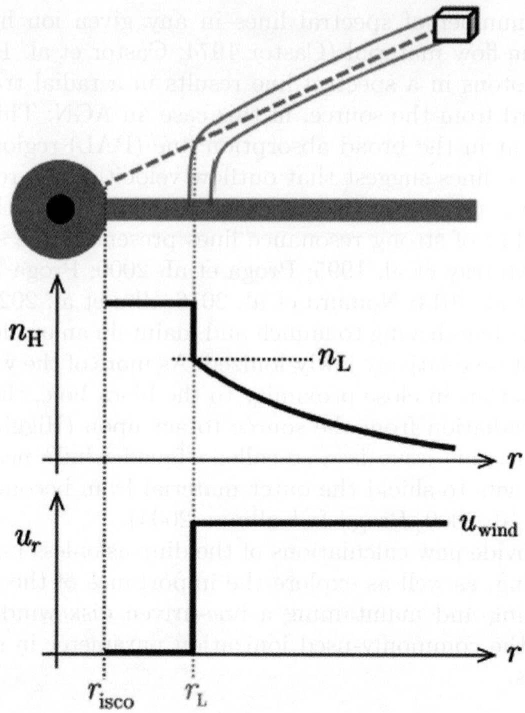

Figure 1. Modeled radial line-of-sight geometry at a given angle above the disk, shown with representative versions of the hydrogen number density n_H and radial flow speed u_r. Dashed line shows the modeled line of sight to the parcel of gas. The radius of the innermost stable circular orbit and launch radius of the wind are denoted by r_{isco} and r_L respectively. The curvature of the flux tube is shown for completeness, but was not considered in this work.

There are five processes that need to be accounted for in this calculation: collisional ionization (CI_i), autoionization (AI_i), photoionization (PI_i), radiative recombination (RR_{i+1}), and dielectric recombination (DI_{i+1}). We do not include three-body recombination, as it generally does not represent a significant contribution to the ionization balance except in the high-density limit, which the AGN environments considered here traditionally do not experience (see, e.g., Rees et al. 1989). The five rates considered are given in units of s^{-1}. For a time steady case, we can write the ionization balance as

$$\frac{n_{i+1}}{n_i} = \frac{CI_i + AI_i + PI_i}{RR_{i+1} + DR_{i+1}}. \tag{1}$$

Using the calculated ionization balance, the luminosity is then recalculated according to the attenuation of the SED due to the intervening material. These calculations are then repeated for each radial step along the line of sight grid.

Because the luminosity is comprised of three separate components, the hot corona, the warm comptonising region, and the cool disk, two limiting cases of the attenuation were considered: attenuation of only the corona component of the luminosity along the line of sight, and equal attenuation of all three luminosity components along the line of sight. These two cases are used to limit the behavior of the force multiplier $M(t)$, where t is a dimensionless optical depth parameter. The full form of $M(t)$ is a measure of the overall strength of the wind at a given optical depth. For both limiting cases discussed above, we found a marked flattening of the $M(t)$ curve, similar to the flattening of the force multiplier shown in Figure 4 of Lattimer & Cranmer (2021).

We additionally calculate values of the ionization parameter ξ and mean ionization number $\mathcal{I}$ for each model. ξ is defined by the hydrogen ionizing luminosity L_X, the distance from the source r, and the hydrogen number density n_H. ξ and $\mathcal{I}$ are given by

$$\xi = \frac{L_X}{n_H r^2} \quad \text{and} \quad \mathcal{I} = \frac{\sum_i (n_{ion}/n_{el})i - 1}{Z}, \tag{2}$$

where Z is the atomic number, i is the ionization state, and n_{ion}/n_{el} represents the fractional abundance of each ion of a given element. From our calculations, we found that multiple values of the mean ionization state of the wind material are possible for the same value of ξ. This suggests that ξ is not a complete description of the ionization state of the wind.

3. Future Work

Lattimer & Cranmer (2021) found that a similarly flattened $M(t)$ curve for a wide range of massive stellar models led to a significant quenching of the stellar wind, as well as a quenching of the wind mass-loss rates. The future of this work involves investigating whether the observed $M(t)$ curve will have a similar affect on the mass-loss rates from the AGN accretion disk. Additionally, work on a self-consistent geometry to represent the attenuation of the three luminosity components is already underway. We also plan to explore alternative methods to the ionization parameter ξ, in order to obtain an more accurate and complete picture of the outflow ionization balance.

References

Castor, J. I. 1974, *MNRAS*, 169, 279.
Castor, J. I., Abbott, D. C., & Klein, R. I. 1975, *ApJ*, 195, 157.
Cavaliere, A., Lapi, A., & Menci, N. 2002, *ApJ* (Letters), 581, L1.
Cunto, W. & Mendoza, C. 1992, *Rev. Mexicana AyA*, 23, 107
Cunto, W., Mendoza, C., Ochsenbein, F., et al. 1993, *A&A*, 275, L5
Del Zanna, G., Dere, K. P., Young, P. R., et al. 2021, *ApJ*, 909, 38.
Dere, K. P., Landi, E., Mason, H. E., et al. 1997, *A&AS*, 125, 149.
Fabian, A. C. 2012, *ARAA*, 50, 455.
Giustini, M. & Proga, D. 2019, *A&A*, 630, A94.
Harrison, C. M., Costa, T., Tadhunter, C. N., et al. 2018, Nature Astronomy, 2, 198.
Higginbottom, N., Proga, D., Knigge, C., et al. 2014, *ApJ*, 789, 19.
Hillier, D. J. 1990, *A&A*, 231, 116
Hillier, D. J. & Miller, D. L. 1998, *ApJ*, 496, 407. doi:10.1086/305350
Hillier, D. J. & Lanz, T. 2001, Spectroscopic Challenges of Photoionized Plasmas, 247, 343
Hopkins, P. F., Torrey, P., Faucher-Giguère, C.-A., et al. 2016, *MNRAS*, 458, 816.
King, A. & Pounds, K. 2015, *ARAA*, 53, 115.
Kramida, A., Yu. Ralchenko, Reader, J., & NIST ASD Team. 2018, NIST Atomic Spectra Database (ver. 5.6.1) [Online]. Available: `https://physics.nist.gov/asd`
Lattimer, A. S. & Cranmer, S. R. 2021, *ApJ*, 910, 48.
Magorrian, J., Tremaine, S., Richstone, D., et al. 1998, *AJ*, 115, 2285.
Mendoza, C., Bautista, M. A., Deprince, J., et al. 2021, Atoms, 9, 12.
Murray, N., Chiang, J., Grossman, S. A., et al. 1995, *ApJ*, 451, 498.
Nomura, M., Ohsuga, K., Takahashi, H. R., et al. 2016, *PASJ*, 68, 16.
Ostriker, J. P., Choi, E., Ciotti, L., et al. 2010, *ApJ*, 722, 642.
Proga, D., Stone, J. M., & Kallman, T. R. 2000, *ApJ*, 543, 686.
Proga, D. 2007, *ASP-CS*, The Central Engine of Active Galactic Nuclei, 373, 267.
Proga, D. & Kallman, T. R. 2004, *ApJ*, 616, 688.
Rees, M. J., Netzer, H., & Ferland, G. J. 1989, *ApJ*, 347, 640.
Risaliti, G. & Elvis, M. 2010, *A&A*, 516, A89.
Silk, J. & Rees, M. J. 1998, *A&A*, 331, L1.
Zhu, Y., Bu, D.-F., Yang, X.-H., et al. 2022, *MNRAS*, 513, 1141.

Black Hole Winds at All Scales
Proceedings IAU Symposium No. 378, 2024
G. Bruni, M. Diaz Trigo, S. Laha & K. Fukumura, eds.
doi:10.1017/S1743921323002879

Study accretion and ejection using a new GPU-accelerated GRMHD code

Asaf Pe'er, Damien Bégué and Guoqiang Zhang
Department of physics, Bar Ilan University, Ramat-Gan, 52900, Israel
email: asaf.peer@biu.ac.il

Abstract. We study disks and jets in various accretion states (SANE and MAD) using novel, GPU-accelerated general-relativistic magneto-hydrodynamic (GR-MHD) code which we developed, based on HARM. This code, written in CUDA-c and uses OpenMP to parallelize multi-GPU setups, allows high resolution simulations of accretion disks and the formation and structure of jets without the need of multi-node supercomputer infrastructure. A 256^3 simulation is well within the reach of an Nvidia DGX-V100 server, with the computation being a factor about 100 times faster if only the CPU was used.

We use this code to examine several disk structures, wind and jet properties in the MAD and SANE states. In the MAD state, we find that the magnetic flux threading the horizon mostly depends on the spin of the BH. This implies that the jet structure and power are strong functions of the spin, with non-spinning BHs have the widest jets.

Keywords. Accretion, Black hole physics, Computational methods, GPU computing, Magneto-hydrdynamical simulations

1. Introduction

Accretion disks around compact objects are central to many astronomical objects of interest, including active galactic nuclei (AGNs), X-ray-emitting binaries (XRBs), and even gamma-ray bursts (GRBs). Theoretically, several types of disks have been identified, whose structures mainly depend on (i) the mass accretion rate and the resulting optical depth and on (ii) the configuration of the magnetic field inside the disk (for reviews, see, e.g., Narayan & McClintock 2008; Abramowicz & Fragile 2013).

Two main types of steady state disk configurations are commonly identified in the literature. They differ by the role of magnetic field in enabling and regulating the accretion. The first is magnetically arrested disks (MAD; see Narayan et al. 2003). In the MAD state, magnetic field lines are accumulated near the horizon, and the resulting magnetic pressure regulates the accretion rate by delaying the accreting material. As a consequence of the strong magnetic field close to the black hole, in the inner parts of the accretion disk, the infalling gas accretes in filaments (Foucart et al. 2017; Wong et al. 2021). This configuration is often accompanied by ejection of matter in the form of strong, relativistic jets. The second type of disk configuration is termed "standard and normal evolution (SANE; Narayan et al. 2012). In this configuration, the magnetic flux threading the horizon is relatively weak, enabling smooth accretion. Magnetic fields therefore contribute mainly to the transport of angular momentum to larger radii. While relativistic jets are observed, they are typically weaker than those seen in MAD state.

In the past two decades, the rapid advance in parallel computation enables to study these complex systems with more and more advanced global numerical simulations. Such simulations are used for time-dependent studies of these complex accreting (and ejecting)

dynamical systems in full general relativity (e.g., using the Kerr metric), shedding light on the magnetohydrodynamical processes in the close vicinity of a black hole (see, e.g., Mizuno 2022 for a recent review). Over the years, several codes solving the general relativistic magnetohydrodynamic (hereinafter GRMHD) equations have been developed (see, e.g., De Villiers et al. 2003; Gammie et al. 2003; Del Zanna et al. 2007; Stone et al. 2008; Porth et al. 2017).

Clearly, such simulations, despite their great success, are very time-consuming and computationally expensive, necessitating access to high-performance computing (HPC) facilities. This naturally provides a physical limitation on the ability to perform such calculations. In the past decade, HPC has seen the development of accelerators (e.g., Giles & Reguly 2014) with the generalization of graphics processing units (GPUs) on computing nodes. Although development of applications for GPUs is a very time consuming task, it is also very rewarding: time-explicit grid-based MHD simulations can effectively run on GPU accelerators, with a large time gain (typically more than an order of magnitude) compared to their CPU counterparts. Today, we are aware of three general-relativistic codes aimed at studying accretion disks with GPU capabilities: grim (Chandra et al. 2017), H-AMR (Liska et al. 2018; Chatterjee et al. 2019),which is based on the hierarchical use of MPI, OpenMP, and CUDA and K-ATHENA (Grete et al. 2021). These codes are not publicly available.

2. cuHARM: an overview

In the past few years we developed a new, GPU-based numerical solver for GRMHD equations in Kerr metric. This code is written in cuda-C, and is based, in many ways, on the publicly available code HARM (Gammie et al. 2003; Noble et al. 2006). We therefore call it cuHARM (cuda-HARM; Bégué et al. 2023). In its current form, cuHARM is designed to run on a single multi-GPU workstation by splitting the volume into equal shares between each GPU. This already allows us to run 3D simulations with resolution of $\sim 192^3$, taking about 72 hr to reach $t = 10^4$ M on an Nvidia DGX- 8xV100 server.

cuHARM solves the ideal, non-resistive GRMHD equations. This system of equations describes the motion of ideal magnetized plasma in arbitrary space-time. The equations that are being solved are the mass, energy, and momentum conservation as well as the homogeneous Maxwell's equations. Here we assume stationary space-time (Kerr metric). These equations characterize the time and space evolution of the gas properties: density, ρ, internal energy density u, four-velocity, u^μ, gas pressure p_g, entropy, S, and magnetic field.

These equations are solved using the finite volume methodology, similar to the original HARM algorithm. We assume that the value of each quantity is uniform within the entire volume of each cell, and calculate the flux of quantities (such as matter, momentum and energy) at the cell's boundaries. The values of the quantities are reconstructed at the boundaries using piecewise linear interpolation method (PLM). In order to satisfy the solenoid condition ($\vec{\nabla} \cdot \vec{B} = 0$), we implement the constrained transport algorithm using the "flux-CT" method of Tóth (2000) and Gammie et al. (2003).

The code is highly optimized, reaching computational efficiency of $\sim 20\%$. With modern GPU workstation we use (Nvidia DGX-V100 server, with 8 V100 GPUs, each with dual Nvlink-2 setup), it updates $\approx 1.5 \times 10^8$ cells per second per card. This results in a typical run time of 2-10 days for resolution $256 \times 256 \times 128$.

Coordinate system. In the results presented below, All the runs were performed using modified, horizon-penetrating Kerr-Schild coordinate system. The grid was refined towards the equator, and cylindrify close to the polar axes (Tchekhovskoy et al. 2011). This configuration allows a for a faster computation by reducing the Courant condition on the time step. Full description of the code appears in Bégué et al. (2023).

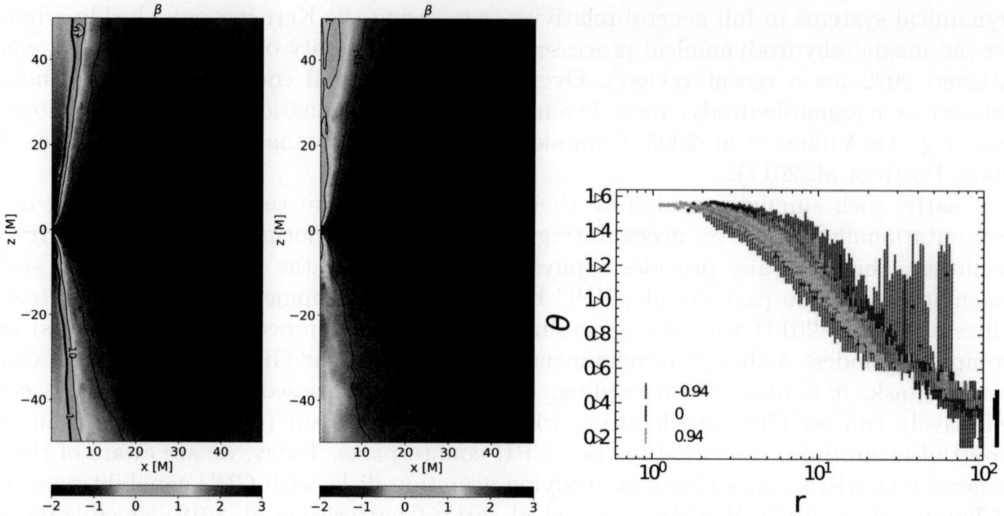

Figure 1. Left. Results of disk/jet in SANE state, showing that when the magnetic flux threading the horizon (left) is high, a jet is formed, while when it drops below a critical value (middle) the polar region is filled with matter, and the jet disappear (from Bégué et al. 2023). Right. The structure of the jet ($r - \theta$) in the MAD state shows widest jet (red) for non-spinning BH. The error bars represent the fluctuations. Figure from Zhang et. al. (in prep.).

3. First results

We performed several runs, both in the MAD and the SANE states. In all runs, we started with initial torus in hydrostatic equilibrium, as described by Fishbone & Moncrief (1976). In this solution, the density scale is not specified, and we normalize it at its maximum value to $\rho = 1$ (McKinney & Gammie 2004). The steady state configuration of the accretion flow (SANE or MAD) is determined by the initial magnetic field configuration, which is set by specifying the magnetic vector potential, A_ϕ (this automatically ensures $\vec{\nabla} \cdot \vec{B} = 0$). Furthermore, the initial intensity of the magnetic field is determined by specifying the maximum value of the gas to magnetic field pressure ratio, $\beta_0 \equiv p_{g,\max}/p_{b,\max} \gg 1$. The initially sub-dominant magnetic field is amplified by the development of magneto-rotational instability (MRI) which enables transport of angular momentum through the disk, and accretion of matter into the black hole.

For runs in the SANE configuration, we find that the accretion rate nearly only depends on the initial mass of the disk. An empirical fit gives $\log \dot{M} = 0.84 \log M - 4.1$, demonstrating that the initial magnetic field and the adiabatic index do not substantially influence the mass accretion rate for the SANE/ RIAF disks considered. The initial magnetization parameter has no significant effect on the accretion rate, with the general trend that the accretion rate weakly increases with β_0. We find that the larger the initial β_0, the faster the spread of the disk. This is to be expected as the Maxwell viscous coefficient scales as $\propto b^r b^\phi$, namely with the magnetic field components. Therefore, larger β_0 leads to a slower initial disk spreading causing a slight increase in the accretion rate. As for the jet production, we find that once the magnetic flux below the horizon drops below a critical value (MAD parameter $\langle \Phi_B(t) \rangle < 1.5$), the magnetic flux threading the black hole is too weak to support the launch of a jet, which disappears. When this occurs, the polar region is filled with low to average density plasma (see Figure 1, left and middle).

Disks in the MAD state are characterized by strong, relativistic jets. For such disks, we explored the dependence of the jet properties on the spin of the central black hole (BH). We found that prograde disks produce wider jets than retrograde disks, and that the widest jets occur for non-spinning BH ($a = 0$; see Figure 1, right). We found that the magnetic pressure dominates the gas pressure in the inner disk regions, $r \lesssim 3r_g$. As the jet boundaries depend on the disk pressure distribution, the magnetic field in the innermost regions is the key to understanding the jet geometry. While the B_ϕ component is the dominant component for most disk configuration, it is found to be negligible compared to B_r at small radii for $a = 0$, which could be associated with the widest jets.

4. Conclusions

We introduce a new, GPU-based global GR-MHD simulation which we call cuHARM. The use of GPUs enable it to speed the calculation by a factor $\gtrsim 20$ relative to similar, CPU-based codes, thereby enabling us both to increase the resolution, as well as explore the evolution of the disk/jet system in a larger parameter space region than previously. Currently, we presented results obtained for disks in the SANE and MAD states. Full description appears in Bégué et al. (2023), as well as in Zhang et. al. (in prep.). As this code is now fully developed, more results are expected in the nearby future.

References

Abramowicz, M. A. & Fragile, P. C. 2013, Foundations of Black Hole Accretion Disk Theory. *Living Reviews in Relativity*, 16, 1.

Bégué, D., Pe'er, A., Zhang, G. Q., Zhang, B. B., & Pevzner, B. 2023, cuHARM: A New GPU-accelerated GRMHD Code and Its Application to ADAF Disks. *ApJS*, 264(2), 32.

Chandra, M., Foucart, F., & Gammie, C. F. 2017, grim: A Flexible, Conservative Scheme for Relativistic Fluid Theories. *ApJ*, 837(1), 92.

Chatterjee, K., Liska, M., Tchekhovskoy, A., & Markoff, S. B. 2019, Accelerating AGN jets to parsec scales using general relativistic MHD simulations. *MNRAS*, 490(2), 2200–2218.

De Villiers, J.-P., Hawley, J. F., & Krolik, J. H. 2003, Magnetically Driven Accretion Flows in the Kerr Metric. I. Models and Overall Structure. *ApJ*, 599(2), 1238–1253.

Del Zanna, L., Zanotti, O., Bucciantini, N., & Londrillo, P. 2007, ECHO: a Eulerian conservative high-order scheme for general relativistic magnetohydrodynamics and magnetodynamics. *A&A*, 473(1), 11–30.

Fishbone, L. G. & Moncrief, V. 1976, Relativistic fluid disks in orbit around Kerr black holes. *ApJ*, 207, 962–976.

Foucart, F., Chandra, M., Gammie, C. F., Quataert, E., & Tchekhovskoy, A. 2017, How important is non-ideal physics in simulations of sub-Eddington accretion on to spinning black holes? *MNRAS*, 470(2), 2240–2252.

Gammie, C. F., McKinney, J. C., & Tóth, G. 2003, HARM: A Numerical Scheme for General Relativistic Magnetohydrodynamics. *ApJ*, 589, 444–457.

Giles, M. B. & Reguly, I. 2014, Trends in high-performance computing for engineering calculations. *Philosophical Transactions of the Royal Society of London Series A*, 372(2022), 20130319–20130319.

Grete, P., Glines, F. W., & O'Shea, B. W. 2021, K-Athena: a performance portable structured grid finite volume magnetohydrodynamics code. *IEEE Transactions on Parallel and Distributed Systems*, 32(1), 85–907.

Liska, M., Hesp, C., Tchekhovskoy, A., Ingram, A., van der Klis, M., & Markoff, S. 2018, Formation of precessing jets by tilted black hole discs in 3D general relativistic MHD simulations. *MNRAS*, 474(1), L81–L85.

McKinney, J. C. & Gammie, C. F. 2004, A Measurement of the Electromagnetic Luminosity of a Kerr Black Hole. *ApJ*, 611, 977–995.

Mizuno, Y. 2022, GRMHD Simulations and Modeling for Jet Formation and Acceleration Region in AGNs. *Universe*, 8(2), 85.

Narayan, R., Igumenshchev, I. V., & Abramowicz, M. A. 2003, Magnetically Arrested Disk: an Energetically Efficient Accretion Flow. *PASJ*, 55, L69–L72.

Narayan, R. & McClintock, J. E. 2008, Advection-dominated accretion and the black hole event horizon. *NewAR*, 51(10-12), 733–751.

Narayan, R., SÄ dowski, A., Penna, R. F., & Kulkarni, A. K. 2012, GRMHD simulations of magnetized advection-dominated accretion on a non-spinning black hole: role of outflows. *MNRAS*, 426, 3241–3259.

Noble, S. C., Gammie, C. F., McKinney, J. C., & Del Zanna, L. 2006, Primitive Variable Solvers for Conservative General Relativistic Magnetohydrodynamics. *ApJ*, 641, 626–637.

Porth, O., Olivares, H., Mizuno, Y., Younsi, Z., Rezzolla, L., Moscibrodzka, M., Falcke, H., & Kramer, M. 2017, The black hole accretion code. *Computational Astrophysics and Cosmology*, 4(1), 1.

Stone, J. M., Gardiner, T. A., Teuben, P., Hawley, J. F., & Simon, J. B. 2008, Athena: A New Code for Astrophysical MHD. *ApJS*, 178(1), 137–177.

Tchekhovskoy, A., Narayan, R., & McKinney, J. C. 2011, Efficient generation of jets from magnetically arrested accretion on a rapidly spinning black hole. *MNRAS*, 418, L79–L83.

Tóth, G. 2000, The ∇· B=0 Constraint in Shock-Capturing Magnetohydrodynamics Codes. *Journal of Computational Physics*, 161(2), 605–652.

Wong, G. N., Du, Y., Prather, B. S., & Gammie, C. F. 2021, The Jet-disk Boundary Layer in Black Hole Accretion. *ApJ*, 914(1), 55.

Author index